Physics Experiments

Grades 3–5

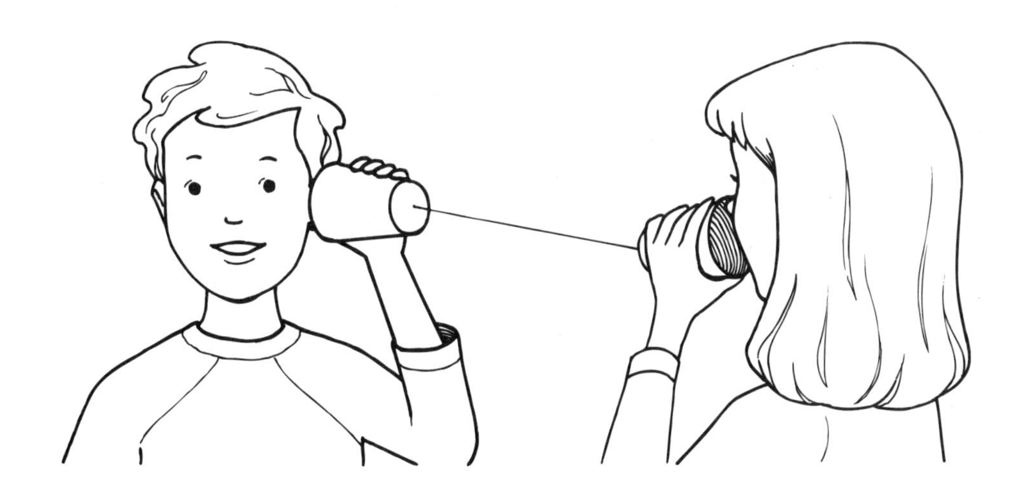

By
Cherie Winner

Illustrations by
Gary Mohrman

Published by Instructional Fair
an imprint of
Frank Schaffer Publications®

Instructional Fair

Author: Cherie Winner

Frank Schaffer Publications®

Instructional Fair is an imprint of Frank Schaffer Publications.

Send all inquiries to:
Frank Schaffer Publications
8720 Orion Place
Columbus, OH 43240-2111

Hands-On Physics Experiments—Grades 3–5

ISBN 0-7424-2749-8

2 3 4 5 6 7 8 9 10 MAZ 10 09 08 07 06

Hands-On Table of Contents

Introduction to This Book 4

Properties of Objects and Materials 5

Experiment 1: Sorting It Out 6

Experiment 2: The Big Drip 8

Experiment 3: How Big? 10

Experiment 4: Water Works 12

Experiment 5: Something's There 14

Experiment 6: Shrinking and Swelling 16

Experiment 7: It's Nothing 18

Experiment 8: Float That Boat 20

Experiment 9: Going, Going, Gone . . . 22

Experiment 10: Rainmaker 24

Further Inquiry Into Properties of Objects and Materials 26

Position and Motion of Objects 27

Experiment 11: Location, Location . . . 28

Experiment 12: Circles in Circles 30

Experiment 13: Getting There 32

Experiment 14: Landmarks 34

Experiment 15: Fast Shadows 36

Experiment 16: Falling Fast 38

Experiment 17: Bouncing Ball 40

Experiment 18: Move It 42

Experiment 19: Pulleys Pull 44

Experiment 20: Short Stack 46

Experiment 21: Bottle Chimes 48

Experiment 22: Bottle Flutes 50

Experiment 23: Getting the Message 52

Further Inquiry Into Position and Motion of Objects 54

Light, Heat, Electricity, and Magnetism 55

Experiment 24: Shadow Shapes 56

Experiment 25: Making Rainbows 58

Experiment 26: Expanding Hands 60

Experiment 27: Light and Heat 62

Experiment 28: Cooling Off 64

Experiment 29: Hot and Cold 66

Experiment 30: Charge It 68

Experiment 31: Series Circuit 70

Experiment 32: Parallel Circuit 72

Experiment 33: Making a Magnet 74

Experiment 34: Pulling Through 76

Experiment 35: Making a Compass . . . 78

Further Inquiry Into Light, Heat, Electricity, and Magnetism 80

Hands-On Introduction to This Book

This book is designed to help your students build on their natural curiosity about the world around them. They will learn how to use the tools and methods of science to look for answers to their questions. The investigations described here support the new National Science Education Standards in their emphasis on science as a process of inquiry rather than a recitation of facts.

Each activity includes an instruction page followed by a student page that can be copied as a handout or used as a transparency. Instruction pages begin with a question or guiding statement that is followed by a materials list, step-by-step directions, suggestions for discussion, and helpful tips. The student pages feature a variety of activities that can be customized to fit the skill level of your students. They include verbal descriptions, drawings, and data found in tables and graphs.

This book is based on the National Science Education Standards for Physical Science. The first section, Properties of Objects and Materials, focuses on what things are made from and other objects that are similar. The second section, Position and Motion of Objects, focuses on where things are and how they move. The final section, Light, Heat, Electricity, and Magnetism, focuses on what we generally think of as *energy*. Feel free to mix activities from different sections. While each experiment stands on its own, you may find that an activity in one section fits well with an activity in another.

At the end of each section, you will find a page of suggestions for further inquiry in that field. Many of these link the Physical Science activities with other portions of the NSE Standards, such as Science and Technology, Earth and Space Science, History and Nature of Science, and Science in Personal and Social Perspectives. All provide additional opportunities for you and your students to explore the fascinating world of physics.

Hands-On Properties of Objects and Materials

In this section your students will classify objects based on the materials from which they are made and the qualities they possess. They will also gain a deeper understanding of the three states of matter—solid, liquid, and gas.

Students will see that similar solids, liquids, and gases have dimension and volume. They will explore ways that water can be used to affect and learn about solid objects. They will make a rudimentary balance that will help them find out that air really is *something*. Their experiments will help them understand the transition from one state to another, as they watch liquid water evaporate to form water vapor and then condense to form water again.

Throughout their explorations, students will gain skill in handling objects and in using measuring devices such as rulers, balances, and thermometers. They will present their results with written descriptions, drawings, arithmetic, and graphs.

"Objects have many observable properties. Those properties can be measured using tools, such as rulers, balances, and thermometers. Objects are made of one or more materials. Objects can be described by the properties of the materials from which they are made, and those properties can be used to separate or sort a group of objects or materials. Materials can exist in different states—solid, liquid, or gas. Some common materials, such as water, can be changed from one state to another by heating or cooling."

National Science Education Standards

Experiment 1 Sorting It Out

Students find similarities and differences among several common objects. Objects can be sorted or separated by various properties.

What You Need:

- Variety of classroom objects (students will find appropriate items)
- Activity sheet on page 7 for each student

What to Do:

1. Discuss how objects can be grouped according to traits they share, such as their function, size, or color. Ask students to come up with other qualities that can be used to sort things.
2. Divide the class into six teams, one for each category listed on the activity sheet. Invite teams to find at least four objects in the room that fit into their category.
3. Compare the lists that are compiled by all the teams. *Could some of the objects fit into more than one category?* Give students time to examine all the objects in all
   the categories and to answer the questions on their activity sheets.

Let's Talk About It:

Objects can be sorted (categorized) by any one of several traits, or they can be sorted by combinations of traits. Such sorting is important in the classification of minerals, chemical elements, and species of plants and animals.

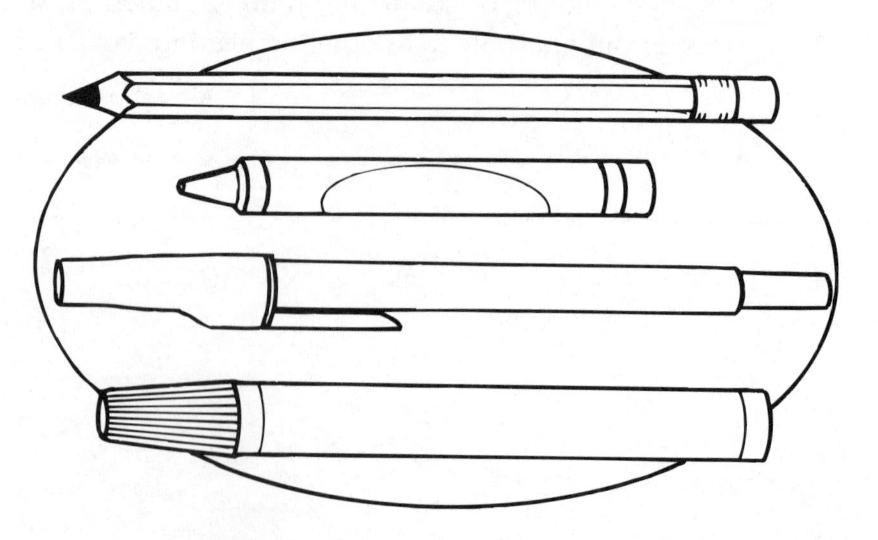

Name ______________________________

Date ______________

Sorting It Out

List the objects your class found that fit into each category.

Containers	Things to write or draw with

Things made partly or completely of wood	Things made partly or completely of plastic

Things that are at least partly green	Things that are at least partly brown

Do some of the objects fit into more than one category? If so, list them in all of the boxes where they belong.

Which object(s) fit into the most categories? ______________________________

____

Did any object(s) fit into just one category? ______________________________

0-7424-2749-8 *Hands-On Physics Experiments*

Experiment 2 The Big Drip

How big is a drop of water?
Water has observable properties.

What You Need:

- A pot or small bucket
- A measuring cup
- Flat pans or trays
- Straws or eyedroppers
- A faucet that can be left dripping overnight
- Activity sheet on page 9 for each student

What to Do:

1. Using straws or eyedroppers, have students experiment with making drops of water on the trays. *How small can they make a single drop?*
2. Open the faucet just enough to create a slight drip. Set the pot or bucket in the sink to catch the drips.
3 Wait at least an hour and then check the bucket. Pour the water into a measuring cup. After measuring it, pour it back into the bucket and continue the experiment.
4. On the next day, you should have students check the bucket again. Measure how much water it contains.

Let's Talk About It:

Over a period of time, a lot of very small drops can add up to a large amount of water.

Teacher Tips:

Use the water you collect to water plants in the classroom or around the school.

Name ______________________________

Date ________________

The Big Drip

Draw a small drop of water and a big drop of water. Make them life size if you can.

small	**big**

After _______ hours of dripping, there were _________ ounces of water in the bucket.

After _______ hours of dripping, there were _________ ounces of water in the bucket.

0-7424-2749-8 *Hands-On Physics Experiments*

Experiment 3 How Big?

Students learn how to find the volume of an odd-shaped object.
Objects have observable properties.

What You Need:

- Modeling clay, a golf ball-sized lump for each group
- Large food can about 2/3 full of water for each group
- Waterproof marker for each group
- Toothpick for each group (wood or plastic)
- Eyedropper
- Small graduated cylinder to measure small amounts of liquid
- Activity sheet on page 11 for each student

What to Do:

1. Have each group mark the starting water level in its can with a line and an *S*.
2. Invite each group to mold its clay into a shape that will fit into the can and does not have thin parts sticking out that might break off during the experiment.
3. Have students stick a toothpick firmly into their clay and use the toothpick as a handle to hold their clay just under the surface of the water. Mark the side of the can to show the new water level. Label this mark *V*.
4. While one team member holds the clay steady at its starting level, have another team member use the eyedropper to lower the water level back to *S*. Pour the extra water into a graduated cylinder and measure it.

Let's Talk About It:

An object will displace an amount of water equal to its own volume. Therefore, the volume of the water you removed will be the same as the volume of the clay you submerged.

Name ______________________

Date ______________

How Big?

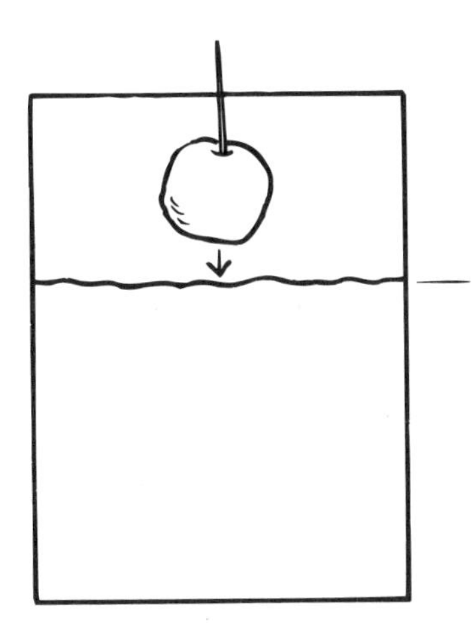

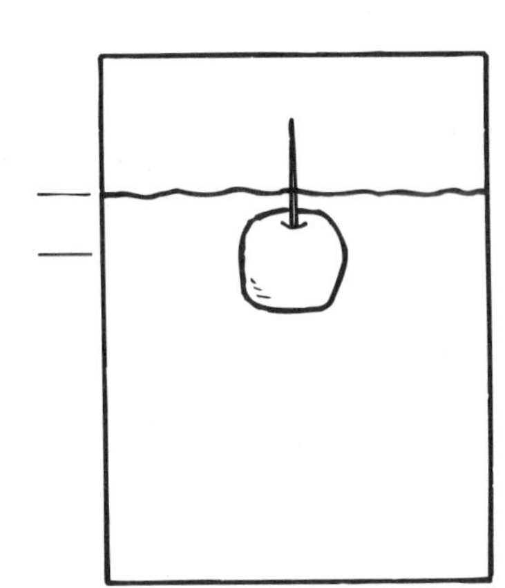

How much water was between *S* and *V*? ______________________

What did your lump of clay look like? Draw it here:

Try this: Refill the water up to level *S*. Change the shape of your clay, and test it again. Is this shape the same volume as the first shape you tried? ________

Experiment 4 Water Works

Do objects weigh the same in water as they do in air?
Water has observable properties.

What You Need:

- Rulers, one for each team
- Pieces of string about one foot long, three for each team
- Two small, non-floating objects for each team; bolts or rocks work well
- Cup or can of water for each team
- Activity sheet on page 13 for each student

What to Do:

1. Help teams to make a balance. Tie one string snugly around the center of a ruler. Hang the ruler from a hook, coat rack, or other stable object. Slide the string along the ruler until the ruler is level.
2. Use the other pieces of string to hang one small object from each end of the ruler. Make the ruler level again by moving the strings along the ruler. At that point, the two objects are balanced.
3. Position the cup of water below one of the balanced objects. Slowly lift the cup toward the object. *What happens when the water touches the object?*

Let's Talk About It:

Because it is more dense than air, water helps support the weight of objects—even objects that do not float. Students may have experienced this themselves while swimming. Under water they can easily lift objects they cannot move on land.

Name ______________________________

Date ________________

Water Works

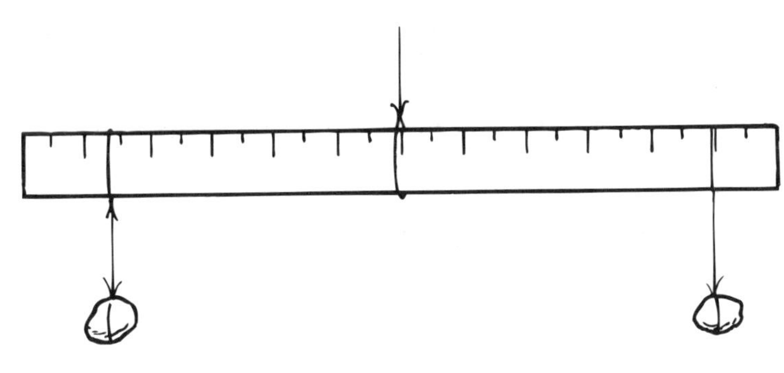

Object #1

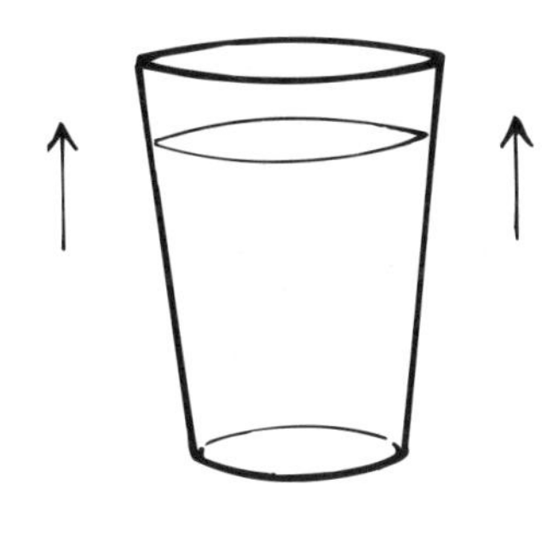

Object #2

Draw what happened when the water touched Object #2.

Experiment 5 Something's There

Does air weigh anything?

Gases have observable properties, including weight.

What You Need:

- Balloons, two of the same size for each student team
- Balances made in previous activity, *Water Works* (page 12)
- Pieces of string about one foot long, two for each team
- Pin (safety or straight)
- Activity sheet on page 15 for each student

What to Do:

1. Have each team blow up its balloons and knot the ends. They then tie a piece of string to each one.
2. Set up the ruler balances again and make them level. Hang one balloon from each end of the ruler. Move the short strings along the ruler to make the ruler level.
4. Pop one balloon with the pin. *What happens?* Wait until the stick stops moving. Pop the other balloon. What happens?

Let's Talk About It:

A balloon full of air weighs more than an empty balloon of the same size. That tells us that air does have mass. It is made of gases whose weight can be measured.

Name ______________________________

Date ________________

Something's There

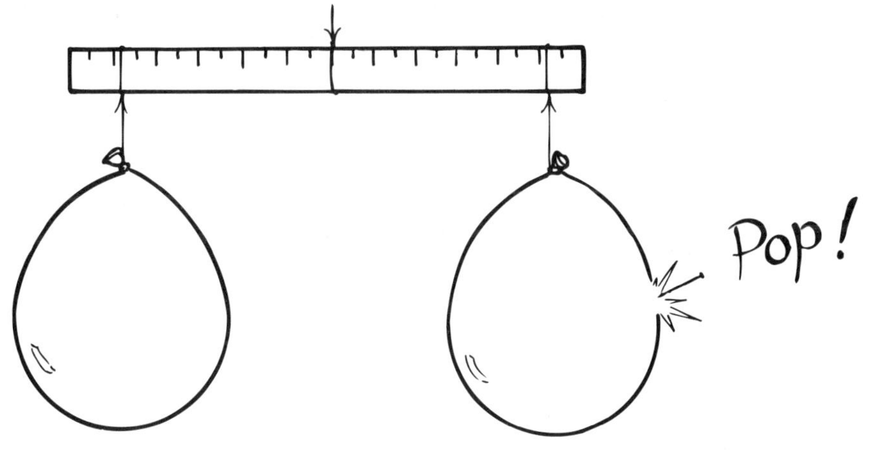

Draw what happened after you popped one balloon.

Draw what happened after you popped both balloons.

0-7424-2749-8 *Hands-On Physics Experiments*

Experiment 6 Shrinking and Swelling

How does temperature affect gases?
Gases expand and contract as temperature rises and falls.

What You Need:

- Balloons, one for each student team
- Markers
- Fabric tape measure
- Three sites with different temperatures: a refrigerator or freezer, a table at room temperature, and a table in the sun
- Three thermometers
- Activity sheet on page 17 for each student

What to Do:

1. Record the temperature at each of your sites.
2. Have students blow up their balloons nearly 3/4 full and tie them closed. Label each group's balloon. Draw a line around each balloon at its widest part. Use the measuring tape to measure each balloon along this line. Record the size.
3. Leave some balloons at room temperature, place some at the warm site, and place the rest at the cold site. Wait at least 20 minutes, and then check the balloons. Quickly measure them along the line again. Record their size.

Let's Talk About It:

The balloons change size as the gases inside them expand (the warm balloons) or contract (the cold balloons).

Name ______________________________

Date ______________

Shrinking and Swelling

Size at the start ______________

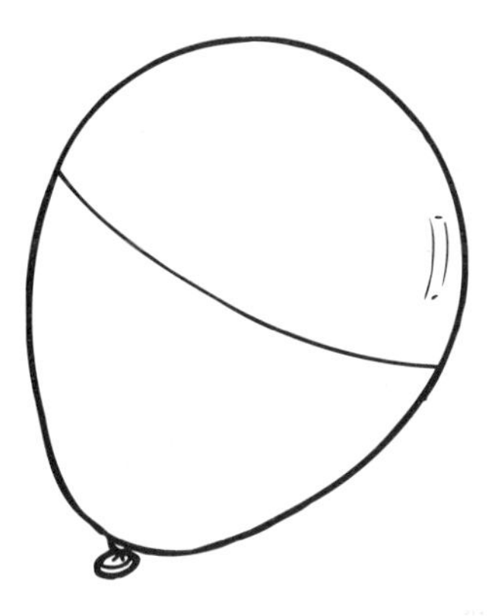

Location of our balloon ____________

Temperature at that site ____________

Size of balloon after 20 minutes ____________

Compare your results with those of other teams:

What happened to the balloons kept at room temperature? ________________

____

What happened to the balloons kept in a warm spot? ____________________

____

What happened to the balloons kept in a cool spot? _____________________

____

____

Do you think the balloons in the cold lost air? Do you think the balloons in the warm spot took in more air? How could you test this?

____

____

Experiment 7 It's Nothing

Students create a partial vacuum.
Gases have observable properties.

What You Need:

- Plastic wrap or thin plastic bags, one for each student team
- Drinking glasses, one for each team
- Rubber bands, one for each team
- Activity sheet on page 19 for each student

What to Do:

1. Push some of the plastic sheet into a glass. Secure the plastic with a rubber band around the top of the glass.
2. Gently pull on the plastic in the glass. *Can you pull it out of the glass?*
3. Insert a pencil under the rubber band and plastic to hold them away from the side of the glass. Try again to pull on the plastic in the glass. *Does it move more easily now?*

Let's Talk About It:

When you pulled on the plastic the first time, you expanded the space available to the air inside the glass. That created a vacuum, which is simply an area of lower pressure. Pulling the plastic toward an area of higher pressure (the air outside the glass) is difficult. When you let air slip in under the rubber band, air pressure inside the glass was about the same as the pressure outside, and the plastic moved easily.

Name ______________________________

Date ________________

It's Nothing

Is it easy or hard to pull out the plastic?

Is it easier or harder to pull out the plastic when you do this?

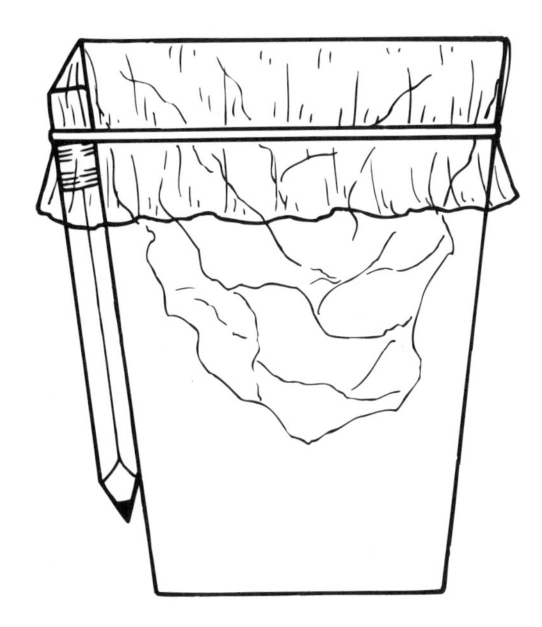

Experiment 8 Float That Boat

Can a sinking object be made to float?

Objects have observable properties because of the materials of which they are made.

What You Need:

- Modeling clay, a golf ball-sized lump for each student team
- Drinking straws
- Scissors
- Bowls or buckets of water or tank the entire class can use
- Activity sheet on page 21 for each student

What to Do:

1. Have students verify that a lump of clay will sink when placed in the water.
2. Invite students to explore how they could make their clay float. Encourage them to record designs that do not work as well as those that do. They must use all their clay in each design. They may add straws or sections of straws (cut with scissors) to their "boats."
3. Share the results. Ask: *Can we reach any conclusions about designs that work or don't work?*

Let's Talk About It:

Students will discover that the shape of their boat has a big effect on its ability to float. They can also explore the possibilities of using straws as outriggers or as internal flotation devices.

Name ______________________________

Date ________________

Float That Boat

Draw a design you tried that did not float.

Why didn't it float?

Draw a design you tried that did float.

Why did this one float?

0-7424-2749-8 *Hands-On Physics Experiments*

Experiment 9 Going, Going, Gone

Does water always evaporate at the same rate?

Water can be changed from one state to another by heating or cooling.

What You Need:

- Narrow, deep containers such as drinking cups, two for each team
- Broad, shallow containers such as oleo tubs, two for each team
- Waterproof markers
- Water
- Measuring spoons
- Two thermometers
- Two sites that will not be disturbed during the experiment, one at normal room temperature, the other much warmer
- Activity sheet on page 23 for each student

What to Do:

1. Label each container with the team's name and where it will be stationed (warm or normal temperature). Measure 4 tablespoons of water into each container.
2. Set deep and shallow containers and a thermometer at each site.
3. On the second day, measure the amount of water remaining in each container. Record your findings. Return the water to its container and set it back at its station.
4. On the third day, measure the amount of water remaining in each container. Fill in the chart on the activity page and plot your results on the graph.

Let's Talk About It:

Water evaporates faster when it is heated and when it has more surface area in contact with the air.

0-7424-2749-8 *Hands-On Physics Experiments*

Name ______________________________

Date ________________

Going, Going, Gone

How much water was present?

	Start	Day 2	Day 3
Room temperature (____°)			
Deep, narrow container			
Shallow, wide container			
High temperature (____°)			
Deep, narrow container			
Shallow, wide container			

Plot your results on this graph.

Use these symbols:

●—● for deep, room temp.

○—○ for deep, high temp.

■—■ for shallow, room temp.

□—□ for shallow, high temp.

Amount of Water (in tablespoons)

4
3
2
1
0

0 | 1 Day | 2 Days

TIME

0-7424-2749-8 *Hands-On Physics Experiments*

Experiment 10 Rainmaker

Students make a model of Earth's water cycle.
Water changes from one state to another by heating and cooling.

What You Need:

- Large jar
- Metal pie tin
- Ice cubes
- Hot water
- Activity sheet on page 25 for each student

What to Do:

1. Pour hot water into the jar. Cover the jar with the pie tin, and fill the tin with ice.
2. Observe what happens in the jar over the next 20–30 minutes.

Let's Talk About It:

The inside of the jar gets cloudy as the water evaporates to become a gas called *water vapor*. This happens in much the same way that clouds form when water evaporates from the surface of the earth. When water vapor reaches the cold pie tin (or the cold upper atmosphere), it condenses—the reverse of evaporation—and becomes water again. You might not actually see rain inside your jar, but students can lift the tin and see water droplets there.

Teacher Tips:

The hotter the water, the better this experiment will work. For safety, *you* handle the hot water.

Name ____________________________

Date ______________

Rainmaker

Draw the changes you see in your model.

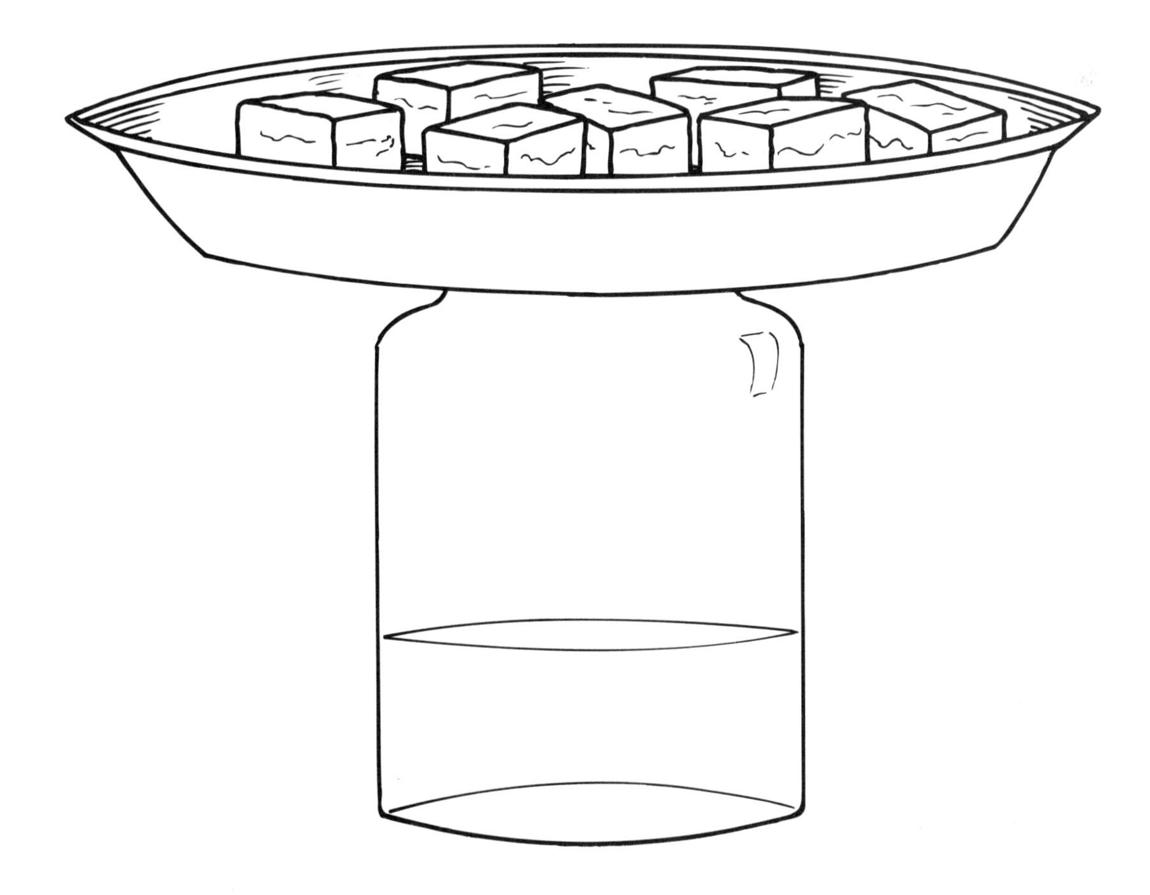

0-7424-2749-8 *Hands-On Physics Experiments*

Further Inquiry Into Properties of Objects and Materials

You can make a game of classifying familiar objects such as writing implements, cooking utensils, books, and sports equipment. (Note that this list already classifies these items!) Help students sort items by their use, the materials from which they are made, their color, their age, and any other criteria your students suggest.

If your students are working on multiplication, have them find the volume of a rectangular object by measuring it in all three dimensions. Then have them find its volume as in *How Big?* (page 10) and compare the results.

After your students have created a partial vacuum in a glass (*It's Nothing*, page 18), have them try the reverse. They secure the plastic to the rim of the glass with the plastic bulging above the glass so there is a lot of air beneath it. Then they try pushing the plastic into the glass. That reduces the space available to the air inside, raising its pressure and making it difficult to push in the plastic.

Discuss real-life examples of using air at low pressure (as in a vacuum cleaner) and at high pressure (as in automobile and bicycle tires).

Point out that the experiment in *Float That Boat*, (page 20) has real-life applications in the design of boats made of metal. Interested students might research how huge ships weighing many tons are able to float. Some colleges host yearly contests in which students design concrete boats. You can also reverse this experiment and try to make a floating object sink.

After collecting the data in *Going, Going, Gone* (page 22), discuss how the students could use their graphs to predict how much water would be left at some other time. How could they make their graphs even more accurate?

The experiments on evaporation (*Going, Going, Gone*, page 22; and *Rainmaker*, page 24) tie in well with units in environmental studies.

Hands-On Position and Motion of Objects

In this section your students will develop their skills of observation, measurement, and perspective. They will encounter key concepts such as gravity, force, and inertia.

First they will describe the position of an object in relation to boundaries of the room. Then your students will describe their own movement along a given path and learn that speed is a measure of change of position over time. They will set objects into motion and examine how fast the objects move and what happens when they strike other objects. They will find that pushing and pulling can be combined to produce greater force, and that pulleys can amplify a person's muscle power.

Students will examine how one object can produce very different sounds, depending on what part of it is made to vibrate. And finally, they will explore the transmission of sound through various materials. All sounds result from vibrations. That is why the National Science Education Standards include Sound in the section on objects in motion. Some teachers prefer to teach sound in their discussions of energy (Section 3). Either way, questions about how sound is created and transmitted make a good link between study units on motion and units on energy.

"The position of an object can be described by locating it relative to another object or the background. An object's motion can be described by tracing and measuring its position over time. The position and motion of an object can be changed by pushing or pulling. The size of the change is related to the strength of the push or pull. Sound is produced by vibrating objects. The pitch of the sound can be varied by changing the rate of vibration."

National Science Education Standards

Experiment 11 Location, Location

Students locate an object relative to its position in the room.
The position of an object can be described by referring to the background.

What You Need:

- Common classroom objects in their usual places
- Yardsticks, one for each student team
- Activity sheet on page 29 for each student

What to Do:

1. Have each team choose an object in the room such as a globe, a terrarium, or a particular desk. Ask them to use a yardstick to measure how far the object is from permanent parts of the room such as the front wall or a window. They then record that information on their activity sheet.
2. For each measurement the students take, have them identify another object that fits the same description. For example, if their object is 3 feet from the back wall, what other objects are also that distance from the back wall?

Let's Talk About It:

Notice that once students find their object's distance from one wall, they do not have to measure how far it is from the opposite wall. Also, measure the vertical dimension. Have students measure their object's distance from the floor.

Position and Motion of Objects

Name ____________________

Date ____________

Location, Location

What object did you locate? ____________________

Describe its location:

This far ____________ from ____________________

What other objects are this same distance ?____________________

____

This far ____________ from ____________________

What other objects are this same distance ?____________________

____

This far ____________ from ____________________

What other objects are this same distance ? ____________________

____

0-7424-2749-8 *Hands-On Physics Experiments*

Experiment 12 Circles in Circles

Students identify objects within a given distance from another object.
The position of an object can be described by reference to other objects.

What You Need:

- String or yarn, about 20 feet for each team
- Yardsticks, one for each team
- Activity sheet on page 31 for each student

What to Do:

1. Have each team use a yardstick and string to create a circle around the object with a radius of about 1 foot. *What objects lie within the 1-foot circle?* (The object may be on a table or shelf, but the circle may be on the floor.)
2. Then have them lay out a circle around the object with a radius of 3 feet. *What objects lie within the 3-foot circle?*

Let's Talk About It:

This activity helps students develop a 360° view, as well as see things from a perspective other than their own location. You may encourage your students to look at what is within a column above and below their object, as well. However, they will have to guess at the exact boundaries of these circles in the air. Notice that everything that is within the inner circle is also contained within the larger circle.

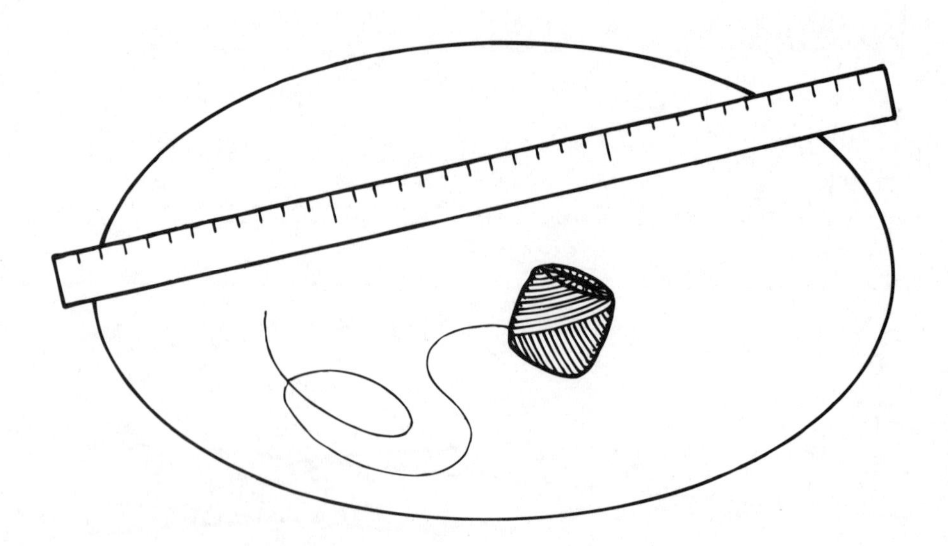

0-7424-2749-8 *Hands-On Physics Experiments*

Name ___________________________

Date ______________

Circles in Circles

List the objects that are inside each circle.

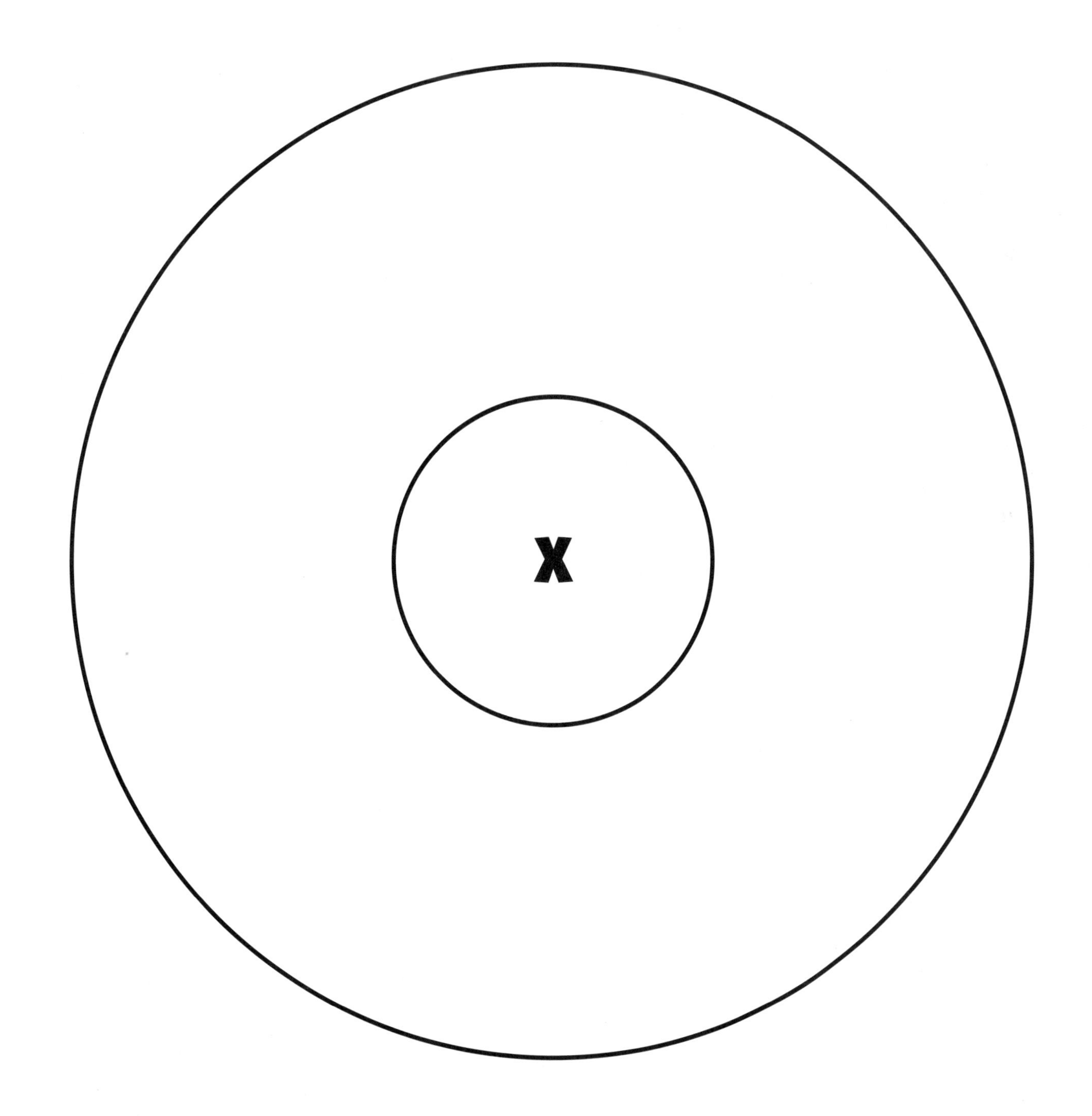

Experiment 13 Getting There

Students map their route to an important place.
Motion can be described by tracing an object's position as it changes over time.

What You Need:

- Students with a sense of adventure
- Activity sheet on page 33 for each student

What to Do:

1. Have each student choose a destination within the room such as the door, a display, or the teacher's desk. That will be their Point *B*. Point *A* is each student's own desk or seat.
2. Ask students to map the route they usually take to get to Point *B* from Point *A*. Have them count the number of steps they take and note all changes in direction. Ask: *Why do you usually go that route?*
3. Now have students map an alternate route to the same Point *B*. Ask: *Why might you choose to go that way?*

Let's Talk About It:

This activity can lead into a fun discussion of why we choose the routes we choose to go anywhere. Some students might always take the shortest route. Others might choose a route that takes them by a friend's desk or a window or that lets them avoid the terrarium with the snake inside.

Name ______________________________

Date ________________

Getting There

Draw your usual route from Point *A* to Point *B*. Show the number of steps you take and where you turn. Now draw a different route to Point *B*. Show how many steps it takes and indicate all turns.

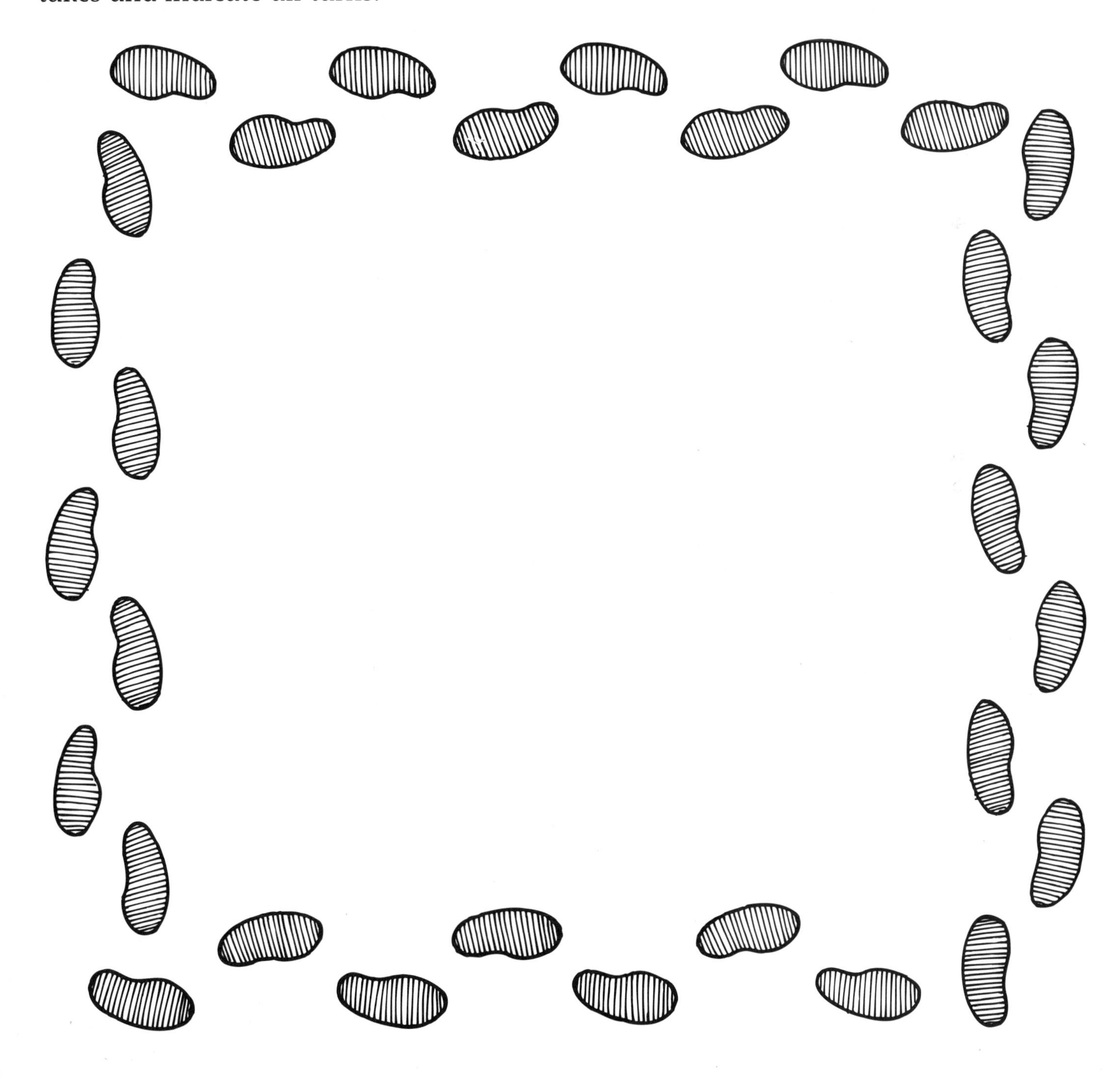

0-7424-2749-8 *Hands-On Physics Experiments*

Experiment 14 Landmarks

Students map a route by using landmarks.
An object's path can be described by tracing its position over time.

What You Need:

- Access to hallways in the school
- Activity sheet on page 35 for each student

What to Do:

1. Invite each team of students to choose a destination within the school. Examples include the cafeteria, library, or the gym. Explain that each team will walk the route and jot down how to reach its destination. Then refer ONLY to directions (such as "Turn left") and landmarks (such as "Third classroom door on the right"). They will not refer to distances, only to objects or structures a person would encounter along the way.
2. When they are farther, have the groups swap directions and see if they can follow the other team's route. Better yet, invite someone from another class to test the directions. If the tester gets confused, find a way to improve the description.

Let's Talk About It:

Using landmarks is a very useful way to give directions, but the same things do not always stand out as *landmarks* to everyone. It can be surprising and informative to try to see your everyday surroundings the way a newcomer might.

Name ______________________________

Date ________________

Landmarks

Write the directions to your destination, using landmarks you see along the way.

____

____

____

____

____

____

____

____

____

____

____

____

____

Did these directions make sense to your tester? ________________

If not, how would you change your directions to make them more clear?

____

____

____

____

Experiment 15 Fast Shadows

How fast does your shadow move?
Speed is the distance traveled in a given unit of time.

What You Need:

- Lamp hung 5–6 feet off the floor
- Room or hallway dark enough that the lamp will cast shadows
- Yardstick
- Watch with a second hand
- Activity sheet on page 37 for each student

What to Do:

1. Ask: *Does your shadow move at the same speed you do? How would you test it?*
2. Mark a "start" line about 10 feet from the lamp and a "finish" line just under the lamp. With one student standing on the start line and the lamp turned ON, mark where the top of the student's shadow is. This will be a few feet beyond the start line.
3. Have a second student keep time as the first student walks slowly from the start line to the finish line. Mark where the top of the walker's shadow is when the walker is at the finish line.
4. Figure out how fast the walker moved and how fast the top of his shadow moved.

Let's Talk About It:

This is a fun way to practice calculating speed rate of movement. It also fits in well with study units on light and shadows.

Teacher Tips:

More advanced students can use their multiplication skills to calculate the speeds in feet per second, kilometers per minute, or any other expression of speed.

Name ________________________________

Date ________________

Fast Shadows

How far did the walker move? ____________

How much time did it take? ____________

How fast did the walker move? $\frac{\textbf{Distance}}{\textbf{Time}}$ =

Finish

How far did the top of the walker's shadow move? ____________

How much time did it take? ____________

How fast did the top of the walker's shadow move? $\frac{\textbf{Distance}}{\textbf{Time}}$ =

0-7424-2749-8 *Hands-On Physics Experiments*

Experiment 16 Falling Fast

Do light and heavy objects fall at the same speed?
Gravity exerts the same force on all objects.

What You Need:

- Quarters, one for each team
- Index cards
- Scissors, one pair for each team
- Activity sheet on page 39 for each student

What to Do:

1. Have each team cut a round piece from an index card. The piece should be a bit smaller than a quarter.
2. Hold the paper disk and the quarter flat and at the same height, about a foot off the floor. Drop them at the same time. Watch closely. *Which hits the floor first?*
3. Now set the paper disk on top of the quarter. Drop them from the same height as before. *Do they separate as they fall, or do they stay together?*

Let's Talk About It:

The disk alone falls more slowly than the coin because it is light enough that friction with the air slows it down. When you put the disk on the coin, they fall at the same rate because the coin "protects" the disk from friction with the air. Gravity causes light and heavy objects to fall at the same speed.

0-7424-2749-8 *Hands-On Physics Experiments*

Name ______________________________

Date ________________

Falling Fast

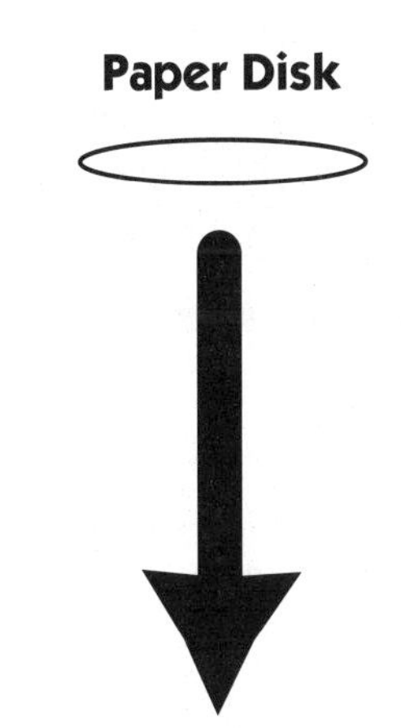

____

Which hits the ground first?

or

Paper Disk and Quarter

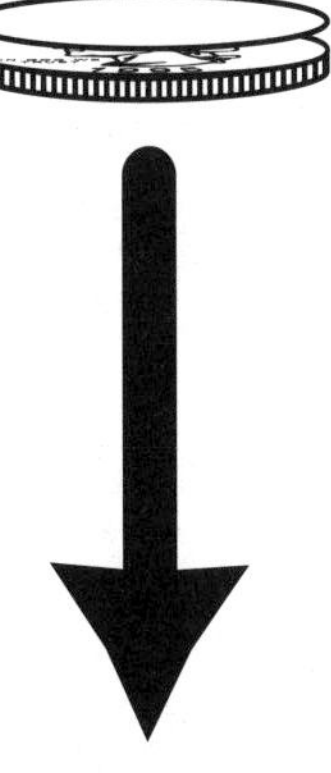

____

What happens?

0-7424-2749-8 *Hands-On Physics Experiments*

Experiment 17 Bouncing Ball

Students combine pushing and pulling forces.
An object's position and motion can be changed by pushing or pulling it.

What You Need:

* Tennis balls, one for each team
* Yardsticks, one for each team
* Hard floor space next to a wall
* Chalk or washable marker
* Activity sheet on page 41 for each student

What to Do:

1. Have each team drop its ball from 1 foot, 2 feet, and then 3 feet. Students mark on the wall how high it goes on its first bounce and measure and record that height.
2. Ask: *How could you make your ball bounce higher?* Have your students push (throw) their ball straight down at the floor, releasing it at about 2 feet above the floor. Can they push with a small amount of force? A greater amount of force? How high does the ball bounce in each case?
3. Clean the marks off the wall. Compare the results.

Let's Talk About It:

When a ball is dropped, gravity pulls it toward the ground. The higher an object is at the start, the longer it will be pulled, the faster it will be going when it hits, and the higher it will bounce. Throwing the ball down adds your push to gravity's pull, increasing the ball's speed and the height of the first bounce.

Name ______________________________

Date ________________

Bouncing Ball

Show how high your ball bounced after hitting the floor once. Did it ever bounce as high as the point from which it was dropped?

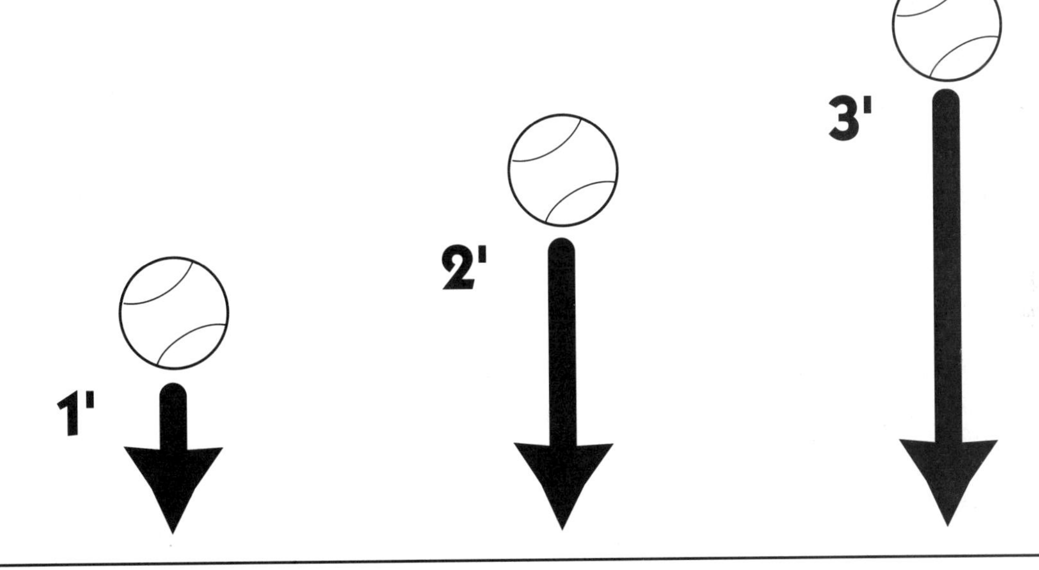

Show how high your ball bounced when you pushed (threw) it from 2 feet high with a small amount of force and then with more force. Did it ever bounce higher than 2 feet?

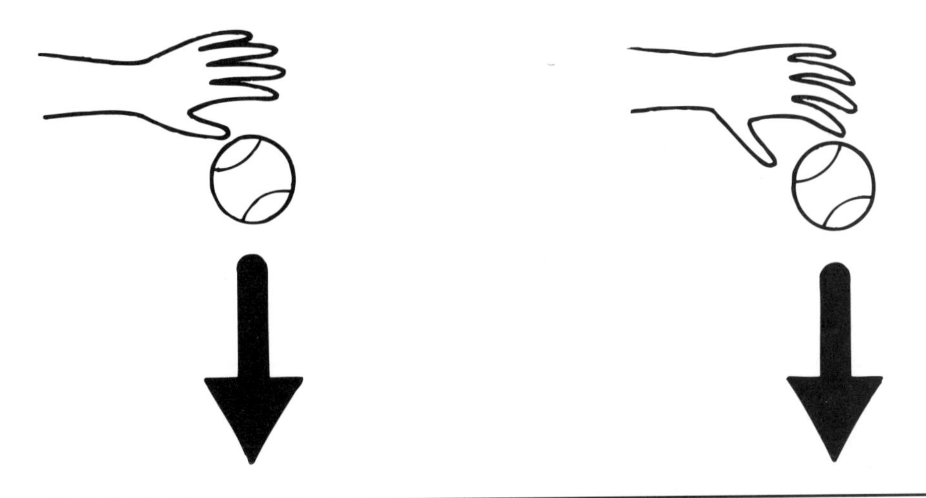

a little force **a lot of force**

Experiment 18 Move It

Is it easier to push or to pull a heavy object?
The position of an object can be changed by pushing or pulling it.

What You Need:

- Two desks or tables 3–4 feet apart with free floor space between them
- Yardstick
- Masking tape
- Activity sheet on page 43 for each student

What to Do:

1. Place strips of tape on the floor to mark the starting positions of the tables. Measure the distance between the tables and mark a line midway between them.
2. With one student behind each table, have them try to push the tables together. Record how far the tables move and whether it was easy or difficult for the students. Put the tables back at the *start* position and add one more student to each side. Keep adding students until the tables can be moved easily.
3. With the tables together at the midway mark, have students try to pull them back to their original position. Again start with one student pulling on each table. Record how far the tables move and how easy or difficult it was to move them. Add a student to each side until moving the tables becomes easy.

Let's Talk About It:

Students will probably find that pulling is easier than pushing. Show them that when they push, some of their force tends to go downward, increasing friction between the table and the floor. When they pull, they usually exert some force upward, thus reducing friction between the table and the floor, so the table slides more easily.

Name ______________________________

Date ______________

Move It

How far could one student push a table? ____________

How many students had to push to make it easy? ____________

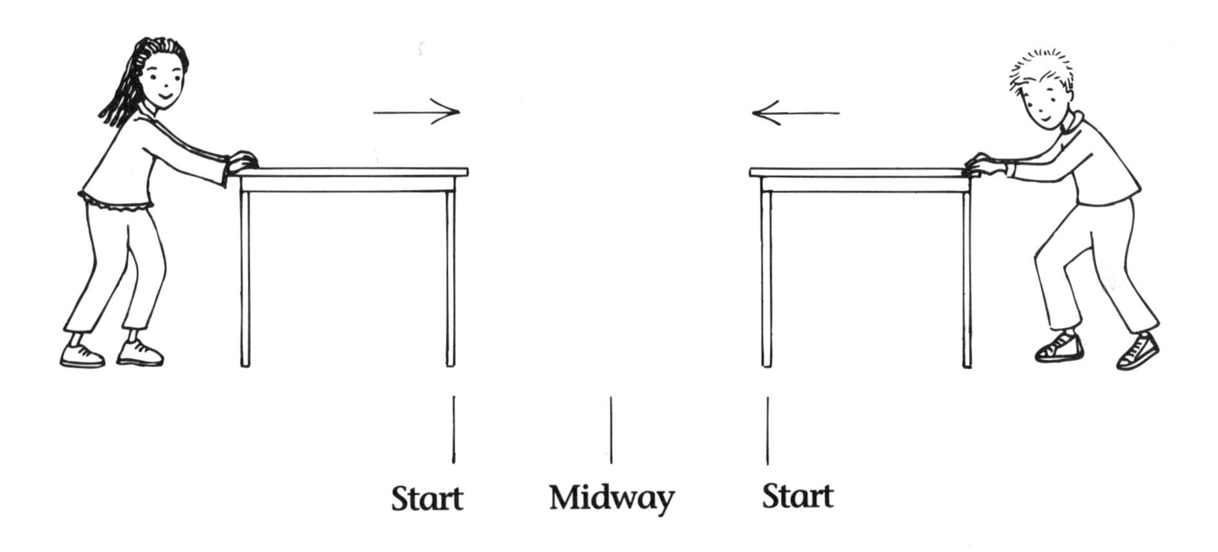

How far could one student pull a table? ____________

How many students had to pull to make it easy? ____________

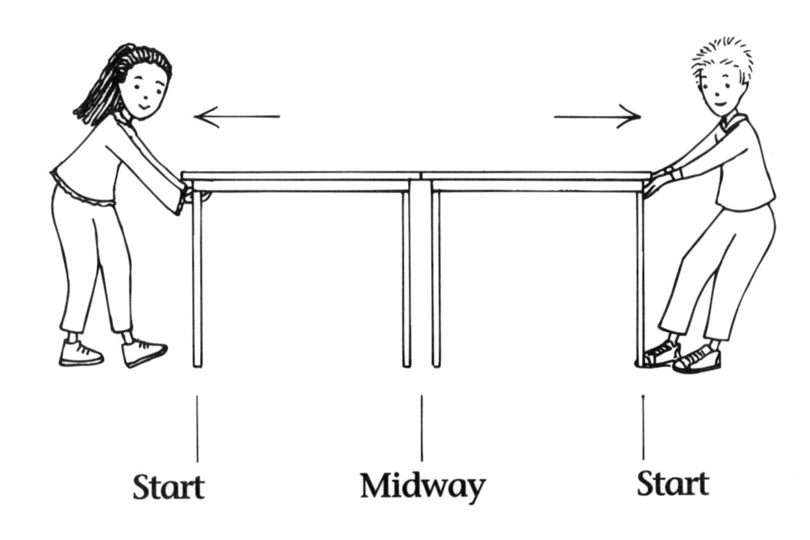

0-7424-2749-8 *Hands-On Physics Experiments*

Experiment 19 Pulleys Pull

Students explore the use of pulleys.
Pulleys magnify the force applied to an object.

What You Need:

- The same tables or desks used in *Move It* on page 42
- Two lengths of rope, each about 30 feet long
- Activity sheet on page 45 for each student

What to Do:

1. Set the tables in their starting positions as in the activity *Move It* on page 42.
2. Tie one rope to a front table leg and the other rope to the other front leg of the same table. Loop each rope around the nearest leg of the second table and back to the first table leg into a figure-eight pattern as shown in the sketch. Run the ropes under the first table so the free ends are behind the table.
3. Have one student pull on the free ends of both ropes. *Do the tables move?*
4. Wrap the rope around the table legs in two more figure eights and try again, this time with one student pulling on the ropes. *What happens?*

Let's Talk About It:

The rope and the table legs act as a *pulley*. A pulley is a simple machine that reduces the amount of force you need to exert to move an object. The more loops you add, the lighter the load feels.

Name ______________________________

Date ________________

Pulleys Pull

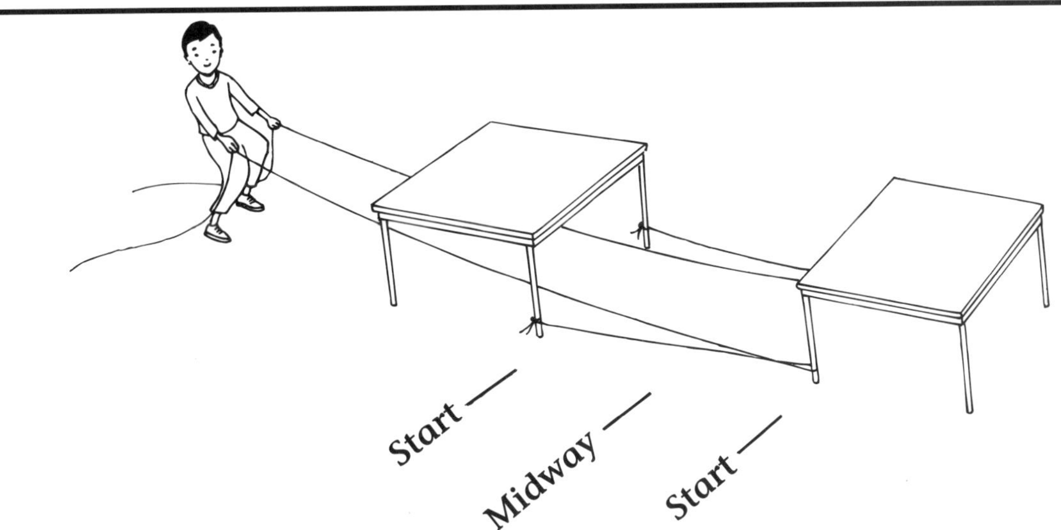

How far could one student pull the tables with the ropes looped once around the legs?

____

Compared to pulling the table directly, was it easier, harder, or about the same?

____

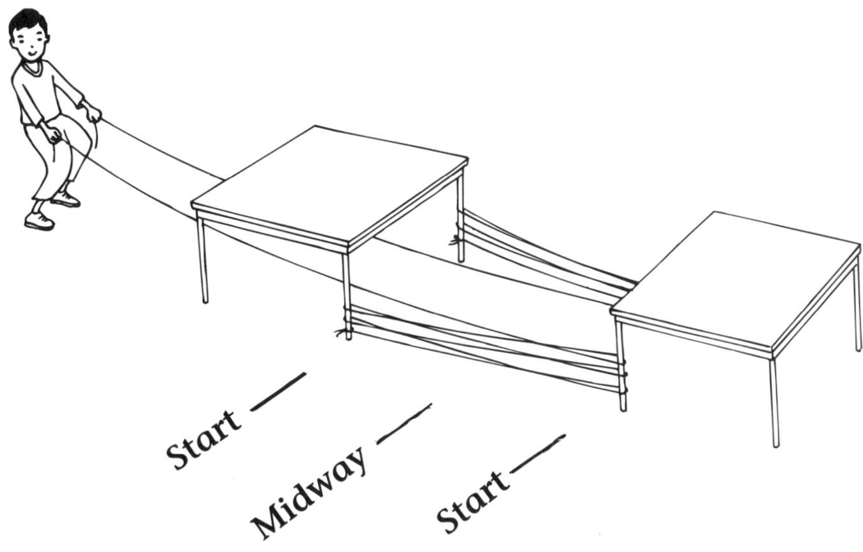

How far could one student pull the tables with the ropes looped three times around the legs?

____

Compared to using just one loop, was it easier, harder, or about the same?

____

Experiment 20 Short Stack

Students see inertia at work.

The strength of an object's motion is related to the strength of the force acting on it.

What You Need:

- Four nickels and one penny for each team
- Activity sheet on page 47 for each student

What to Do:

1. Have students stack their nickels on the marked circle on the activity sheet and place the penny on the other circle. Show them how to flick the penny toward the stack of nickels. Let each student practice this movement. *What happens when the penny hits the nickels very lightly? What happens when it hits the nickels firmly?*
2. Invite students to try this with shorter and taller stacks of pennies. Teams can pool their nickels to make taller stacks. *Does it still work?*

Let's Talk About It:

Since the penny is thinner than a nickel, it strikes only the bottom nickel. When hit with a small amount of force, the nickels do not move. With more force, the bottom nickel moves horizontally, but the nickels above it fall straight down. Both results are because of *inertia,* which is the tendency of a stationary object to remain in place unless a big enough force acts on it. Inertia also means an object in motion will tend to keep moving the same direction, unless an external force acts on it. Your students might find that a taller stack falls straight down more reliably than a shorter stack. The movement of the bottom nickel can be enough to nudge aside one or two nickels above it, but will not be forceful enough to move a tall stack.

Name ______________________________

Date ________________

Short Stack

Stack your nickels here.

Shoot your penny from here.

Experiment 21 Bottle Chimes

How can you change the sound made by tapping on a bottle?
The pitch of a sound can be varied by changing the rate of vibration.

What You Need:

- Three glass bottles of the same size with narrow necks; soda or vinegar bottles work well
- Water
- Hard plastic ruler
- Activity sheet on page 49 for each student

What to Do:

1. Pour different amounts of water into each bottle.
2. Tap on each bottle with the edge of the ruler.
   Which bottle makes the lowest sound? Which makes the highest sound?
3. Grip each bottle tightly as you tap on it again.
   Does that change the sound?

Let's Talk About It:

When you tap a bottle, the glass vibrates and produces a bell-like sound. Slower vibrations create lower notes. Water in the bottles slows the rate of vibration. The more water in the bottle, the slower the glass will vibrate and the lower the pitch will be. Gripping the bottle reduces the vibration of the glass and dampens the sound without changing the pitch.

Teacher Tips:

You may find books that say tapping a bottle with more water in it will make a higher note. Try it for yourself and find out. Then go on to the next activity and find out what happens when it is the air inside the bottle, rather than the glass, that creates the sound.

Name ______________________________

Date ________________

Bottle Chimes

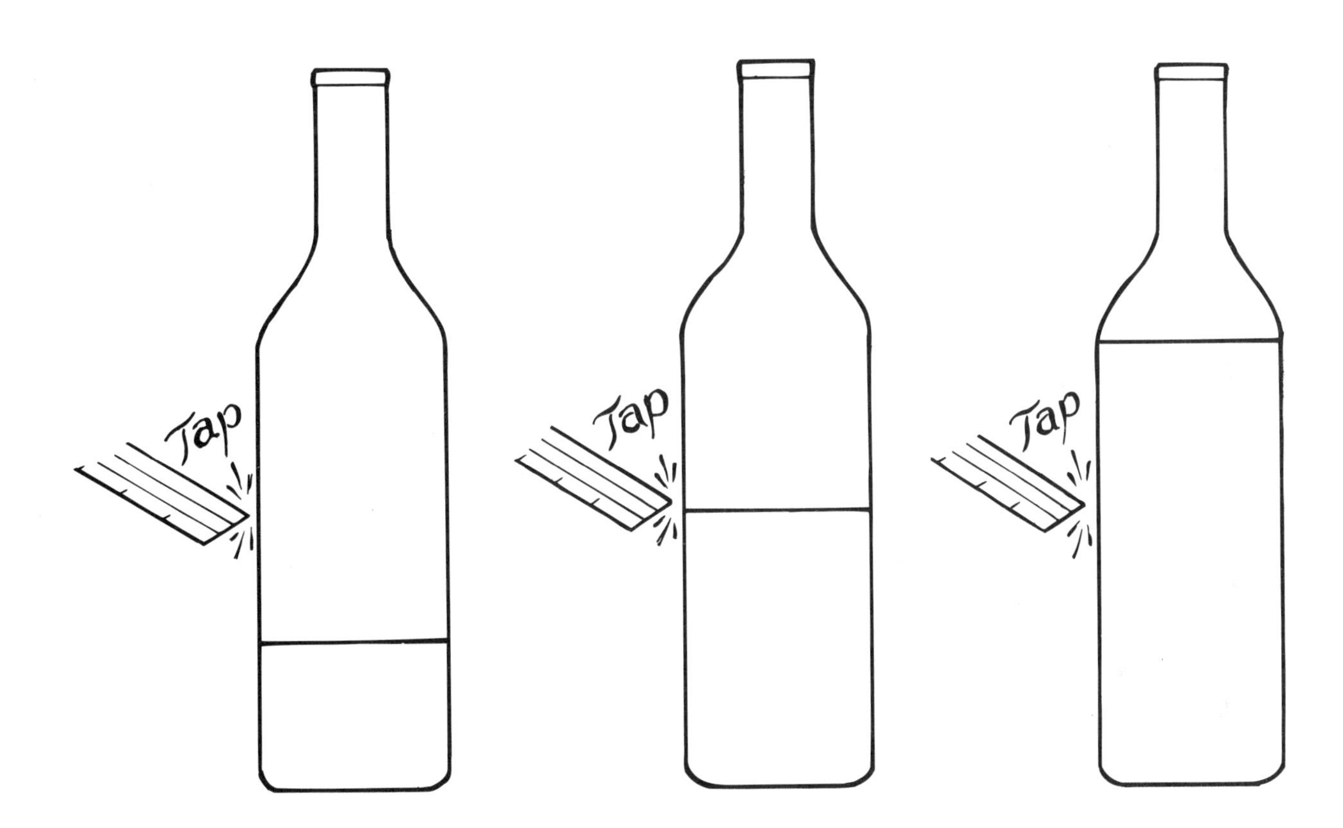

Which bottle chime makes the highest note? ______________________________

Which makes the lowest? ______________________________

What vibrates to produce the notes you hear? ______________________________

____

What happens when you grip each bottle as you tap on it?

____

____

0-7424-2749-8 *Hands-On Physics Experiments*

Experiment 22 Bottle Flutes

How can you change the sound made by blowing into a bottle?
The pitch of a sound can be varied by changing the rate of vibration.

What You Need:

- The three bottles containing different amounts of water that you used in *Bottle Chimes* on page 48
- Activity sheet on page 51 for each student

What to Do:

1. Blow down an inside edge of each bottle. What do you hear? Which bottle makes the lowest sound? Which makes the highest sound?
2. Grip each bottle tightly as you blow into it again. Does that change the sound?

Let's Talk About It:

When you blow into a bottle, the air inside vibrates and produces a tone. The more water there is in the bottle, the shorter the column of air above it, and the higher the note when the air vibrates. Gripping the bottle does not affect the air inside or the sound it produces.

Teacher Tips:

This activity and the previous one present an intriguing puzzle. You might do them first as a demonstration, and then ask your students to experiment with the bottles themselves and try to figure out why tapping and blowing produce different results.

Name ____________________________________

Date ____________________

Bottle Flutes

Which bottle flute makes the highest note? ________________________________

Which makes the lowest? ____

What vibrates to produce the notes you hear? ______________________________

____

What happens when you grip each bottle as you blow into it? __________________

____

0-7424-2749-8 *Hands-On Physics Experiments*

Experiment 23 Getting the Message

Students make and test earphones.
Sound is produced by and transmitted through materials that vibrate.

What You Need:

- Two paper cups for each pair of students
- String
- Scissors
- Paper clips
- Activity sheet on page 53 for each student

What to Do:

1. Show students how to make a set of earphones. Poke a hole in the bottom of each paper cup. Slip the end of a string through the hole from outside to inside and tie it to a paper clip inside the cup. Do the same with the other end of the string and the second cup. Have teams use different lengths of string. The shortest string should be 3 feet long.
2. Students then test their earphones. One student will hold a cup to one ear while the partner talks or hums into the other cup. *What does the listener hear if the string connecting the earphones is taut? What is heard if someone pinches the taut string? What if the string is allowed to sag?*

Let's Talk About It:

Sound is transmitted by vibrations. Often, vibrations of solid objects create a more vivid sound than those coming to us through the air. A taut string transmits vibrations clearly, but if the string goes loose, it cannot vibrate in a regular way and the sound is lost. Likewise, pinching the string will prevent sound waves from traveling along the string to the listener's ear.

Name ______________________

Date ______________

Getting the Message

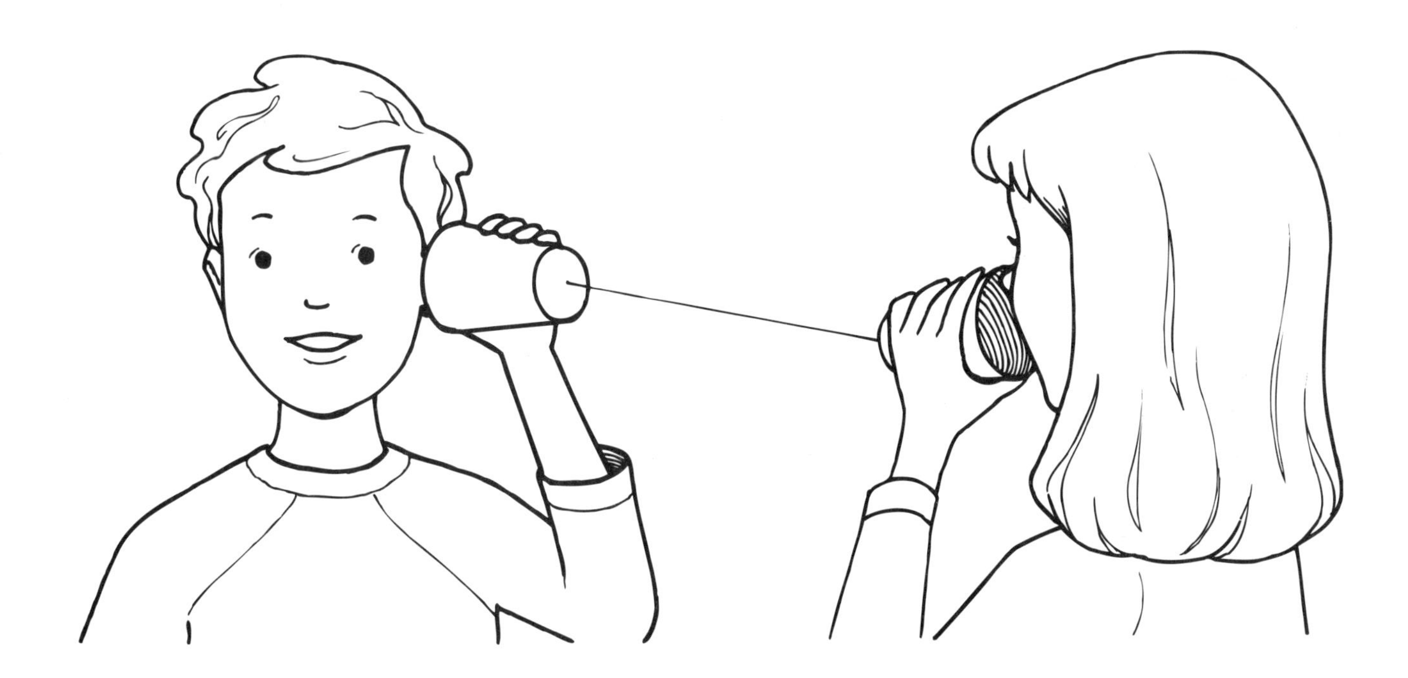

How does the sound change

if you cover the ear that does not have an earphone? ______________________

____

if you keep the string taut, but hold the earphone away from your ear? __________

____

if you pinch the string while your partner hums? ______________________

____

if you let the string go loose? ______________________

____

0-7424-2749-8 *Hands-On Physics Experiments*

Further Inquiry Into Position and Motion of Objects

A fun variation on the *Location, Location* (page 28) is to have two different classes do the activity and then swap rooms and descriptions. *Can they find each other's objects based only on the measurements provided?*

The question of why paths and roads are where they are is a rich area for further exploration. Your students can survey paths across their playground or a local park. They can use both measurements and landmarks to map their route between home and school or other destinations important to them.

After students have measured the speed of their own shadows (*Fast Shadows*, page 36), they can measure the speed of a flagpole's shadow. *Does it shrink as fast in the morning as it does in the afternoon? Does the flagpole's shadow move at the same speed in all seasons of the year?*

If your students are doing multiplication and division, *Bouncing Ball* (page 40) provides the opportunity to practice both. Have students calculate the average height of the first bounce from each height. This is also a good place to introduce more advanced students to the concepts of potential and kinetic energy.

Discuss with your students whether pushing or pulling is generally easier for them. Describe the difference between rear-wheel drive and front-wheel drive. *Which one pushes the car, and which one pulls the car? Which is more efficient?*

If you have access to musical instruments, ask students to identify which ones use vibrating air to make their tones and which use a vibrating solid. How is the pitch changed on the various instruments?

You can explore how sound is transmitted through various solids such as a tabletop or the ground and through liquids such as water. *For example, if you knock two blocks together, what sound do they make if both you and they are in the air? If one of you is underwater? If both are underwater?*

Light, Heat, Electricity, and Magnetism

In this section your students will explore many forms of energy and see how various forms of energy are related to each other.

Students will watch as a beam of light strikes a solid object at different angles or is refracted through lenses made of glass or water. They will look closely at the colors in a rainbow they make themselves and will find an easy way to magnify objects. They will study how color influences the transformation of light into heat and examine how fast heat is lost to cooler surroundings.

Students will find that static electricity can be generated "out of thin air." They will expand their understanding of electric circuits by building series and parallel circuits with a battery and light bulbs. They will also use an electric current to create a magnet.

Finally, they will explore some of the remarkable properties of magnets.

During their work in this section, students will develop skill in handling sources of heat, light, and electricity. Safety is a key concern here. Be certain that your students know that energy in all its forms must be handled with care. In particular, water and electricity are a dangerous combination.

"Light travels in a straight line until it strikes an object. Light can be reflected by a mirror, refracted by a lens, or absorbed by an object. Heat can be produced in many ways . . . . Heat can move from one object to another by conduction. Electricity in circuits can produce light, heat, sound, and magnetic effects . . . . Magnets attract and repel each other and certain kinds of other materials."

National Science Education Standards

Experiment 24 Shadow Shapes

Students find that one object can cast many different shadows.
Light travels in a straight line until it strikes an object.

What You Need:

- Common classroom or household objects
- Desk lamps or flashlights, one for each student team
- Room darkened enough that the lamps will create visible shadows
- Activity sheet on page 57 for each student

What to Do:

1. Invite each team to choose an object from your collection.
2. Have students explore the many shadows cast by their object, using a lamp or flashlight as the source of light. Ask: *Do all of the shadows of your object look like the real object? Why does the shadow change shape?* Give students time to draw the shadows they find most interesting.

Let's Talk About It:

The shape of the shadow depends on where the light source is and how the object is oriented.

Name ______________________________

Date ______________

What object did your team use? ______________________________

Draw two different shadows cast by your object.

Shadow 1

Shadow 2

Experiment 25 Making Rainbows

Students make miniature rainbows.
Light can be refracted by a lens and split into its component colors.

What You Need:

- Tall drinking glasses made of glass, one for each team
- Water
- Sunny day
- Activity sheet on page 59 for each student

What to Do:

1. Have students fill each glass about 3/4 full of water. They then take their glasses and their activity pages to a sunny part of the room.
2. When students hold a glass above the paper, sunlight passing through it forms patterns of light on the page. Some parts of the pattern will be *white* light and other parts will have color. Give students time to experiment with making these little rainbows. Suggest they try holding the glass at different heights, tilting it, and turning it.
3. Ask: *What colors do you see? Are they always in the same order?*

Let's Talk About It:

Each wavelength of light produces a different color. When blended together, as in daylight, the light looks colorless to us. A glass of water acts as a lens that refracts or bends the light waves. At certain angles, different wavelengths bend different amounts. They then separate into distinct bands as in a rainbow, and we can see the different colors.

Teacher Tips:

Be certain your students draw what they actually see. The order of colors—from shortest wavelength to longest—is red-orange-yellow-green-blue-violet. Often one or two bands will appear much broader than the others.

Name ______________________________

Date ______________

Making Rainbows

Draw one of the miniature rainbows made by your glass of water. Show the bands of different colors. What shape were they? How broad were they?

0-7424-2749-8 *Hands-On Physics Experiments*

Experiment 26 Expanding Hands

Students make a magnifying lens.
Water refracts light in a way that makes objects appear larger.

What You Need:

- Sealable clear plastic sandwich bags, one for each team
- Water
- Activity sheet on page 61 for each student

What to Do:

1. Have each group fill its bag with water and seal it completely.
2. As they hold their bag in front of them with one hand, have them move their other hand from beside or beneath the bag to behind it. Does their hand look different when seen through the bag of water?

Let's Talk About It:

The bag of water acts as a magnifying lens. It bends the light waves passing through it in such a way that it creates a larger image than we see when the waves pass through air. Students will find the greatest magnification with their hand a few inches behind the bag.

Teacher Tips:

This is a fun experiment but potentially messy. Do it over a sink, a bucket, or a table covered with towels.

0-7424-2749-8 *Hands-On Physics Experiments*

Light, Heat, Electricity, and Magnetism

Name ______________________________

Date ______________

Expanding Hands

Trace your hand here:

Draw how big your hand looked behind the water lens:

Experiment 27 Light and Heat

Do dark and light colors absorb the same amount of solar energy?
Heat can be produced by the absorption of light.

What You Need:

- Two cups, mugs, or bowls per team. Each team's containers should be of similar size, shape, and material, but should be a different color. One should be black or very dark. The other should be white or pale.
- Two small thermometers for each team
- Two sites, a sunny windowsill or tabletop and a shaded one
- Activity sheet on page 63 for each student

What to Do:

1. Assign half of the teams to the shaded site and half to the sunny site. Have each team invert its containers at its site and place a thermometer under each one.
2. After an hour, touch the containers. *Which one feels warmer?* Lift the containers and record the temperature under each one. Share the results.

Let's Talk About It:

Dark colors absorb more light, and that raises their temperature. Light colors reflect more light and do not gain as much heat. Containers in the shade will stay at room temperature. This experiment shows the link between two forms of energy—light and heat.

Name ______________________________

Date ________________

Light and Heat

Sunny site

________°F   Temperature under each container   ________°F

Shaded site

________°F   Temperature under each container   ________°F

Experiment 28 Cooling Off

How fast does warm water cool down?

Heat moves from areas of higher temperature to areas of lower temperature.

What You Need:

- Two identical cups, bowls, or other small containers for each team
- Warm water (warmer than room temperature)
- Hot water (much warmer than room temperature)
- Two thermometers for each team
- Activity sheet on page 65 for each student

What to Do:

1. Have each team pour warm water into one container and the same amount of hot water in the other container. Place a thermometer in each. After a minute or so, record the starting temperature. Set the containers where their surroundings are the same temperature. (Do not put one in the sun and one in the shade.)
2. Record the temperature of each sample every five minutes. Continue for a half hour.
3. Help your students graph their results. Ask: *How much heat did the hot sample lose in 5 minutes? 10? 20? How much heat did the warm sample lose in the same time?*

Let's Talk About It:

Heat flows from areas of higher temperature to areas of lower temperature. The greater the difference between the two areas, the faster the heat will move.

Teacher Tips:

If you do not have enough thermometers to go around, divide the class into two teams. Have each team track the temperature change in one container.

0-7424-2749-8 *Hands-On Physics Experiments*

Name ______________________________

Date ______________

Cooling Off

Temperature at:

	Start 5 Minutes	10 Minutes	15 Minutes	20 Minutes	25 Minutes	30 Minutes
Hot Water						
Warm Water						

What was room temperature? __________

Graph your results here:

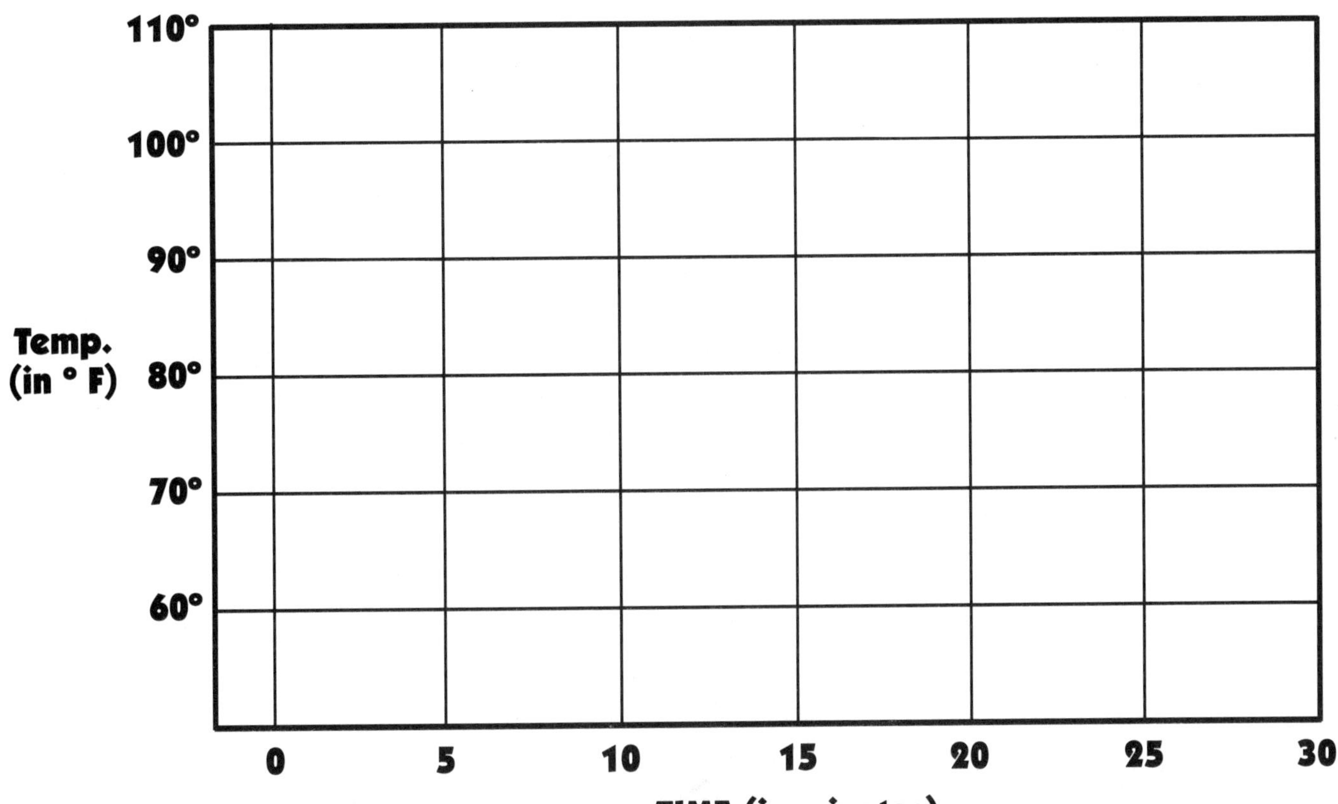

0-7424-2749-8 *Hands-On Physics Experiments*

Experiment 29 Hot and Cold

Does insulating material work for both hot things and cold things?
Heat can move from one object to another by conduction.

What You Need:

- Plastic foam cups, two for each student team
- Hot water
- Cold water
- Thermometers, one for each cup, plus one to register the room temperature
- Clock or watch
- Activity sheet on page 67 for each student

What to Do:

1. Set the cups on a table where they are not disturbed for 30 minutes. Be certain all parts of the table are at about the same temperature.
2. Have students label their cups with their team name and an *H* or a *C*. Pour hot water into the *H* cup and the same amount of cold water into the *C* cup. Have students record the temperature of each of their samples. Leave the thermometers in the water.
3. Check the temperatures again after 10, 20, and 30 minutes.
4. Compare results. How much did the temperature of each sample change? Did the foam cups keep the hot water hot and the cold water cold?

Let's Talk About It:

Since heat flows from areas of higher temperature to areas of lower temperature, the hot water will lose heat to its surroundings and the cold water will gain heat from its surroundings. An insulating material such as plastic foam reduces the movement of heat in either direction.

Name ______________________

Date ______________

Hot and Cold

What was the temperature of the room during the experiment? ______

	Cold Water	Hot Water
Temperature of sample at start		
Difference between sample and room temperature at start		
Temperature of sample at 10 minutes		
Temperature of sample at 20 minutes		
Temperature of sample at 30 minutes		
Change in temperature of sample from 0–30 minutes		
Difference between sample and room temperature at 30 minutes		

Experiment 30 Charge It

Students experiment with static electricity.
Electricity can exist in objects.

What You Need:

- Two balloons for each team, inflated and tied off
- Pieces of wool cloth (a sock or sweater works well)
- Small plate with salt sprinkled on top
- Water tap
- Activity sheet on page 69 for each student

What to Do:

1. Have teams rub one balloon with the wool cloth. Students should rub many times but not too hard, as the balloon might pop. The other balloon is an uncharged control.
2. Invite students to test their balloons in various ways. *If they hold a balloon near their face, do they feel the charge of the balloon? Does the balloon make their hair move? If they hold a balloon near a stream of running water, what does the water do?* (Note: If the balloon touches the water, it will lose its charge and will have to be recharged.) *If they hold a balloon just above the salt, what happens?*

Let's Talk About It:

Rubbing the balloons with wool transfers electrons to them. That gives them a negative charge and the ability to attract materials with a neutral or positive charge, and to repel other negatively-charged objects. We experience this charge as *static electricity*. *Static* here means stationary, in contrast to *current* electricity that flows through circuits like the water current in a stream.

0-7424-2749-8 *Hands-On Physics Experiments*

Name ______________________________

Date ______________

Charge It

Show what happened.

Balloon rubbed with wool

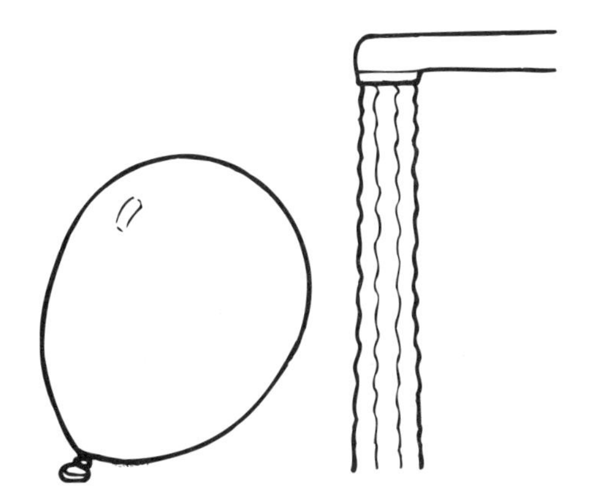

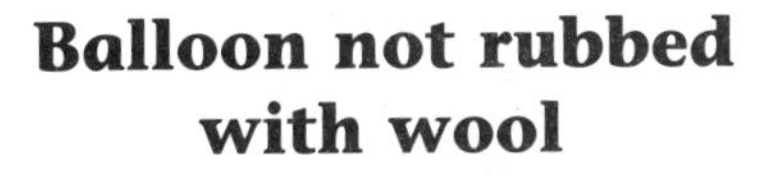

Balloon not rubbed with wool

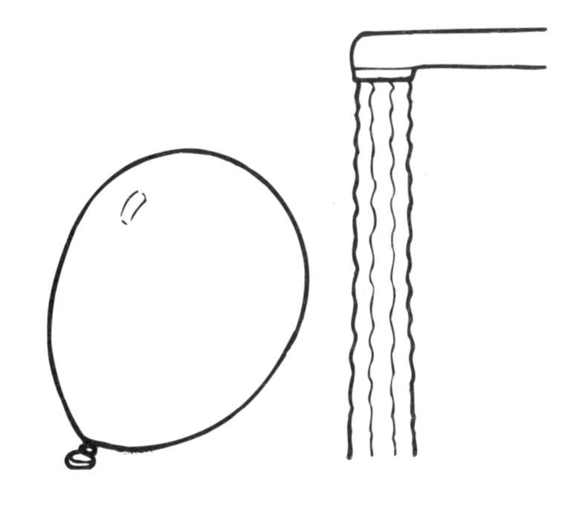

0-7424-2749-8 *Hands-On Physics Experiments*

Experiment 31 Series Circuit

What happens when an electrical circuit runs through more than one light bulb? Electricity in circuits can produce light.

What You Need:

- Six-volt battery for each team
- Two 6- or 12-volt light bulbs in sockets for each team
- Insulated wire with the ends stripped, one 2-foot piece for each team
- Insulated wire with the ends stripped, two 1-foot pieces for each team
- Activity sheet on page 71 for each student

What to Do:

1. Have each team build a circuit that includes the battery and one light bulb in a socket, as shown in the top sketch on the activity sheet. *When both battery terminals are hooked up, what happens?*
2. Have the teams disconnect the long wire at both ends and attach one end of it to the second socket. Then use a short wire to connect the other screw of the second socket to the empty screw of the first socket. *When they touch the free end of the long wire to the free battery terminal, what happens?*
3. Leave all the wires connected as in step #2, but unscrew one of the bulbs. *What happens?* Screw the bulb back in. *What happens?*

Let's Talk About It:

In a series circuit, current runs through two or more appliances (the bulbs) before returning to the battery. Removing a bulb, like a switch being turned off, breaks the circuit.

0-7424-2749-8 *Hands-On Physics Experiments*

Name ______________________________

Date ________________

Series Circuit

Draw colored lines to show the current.

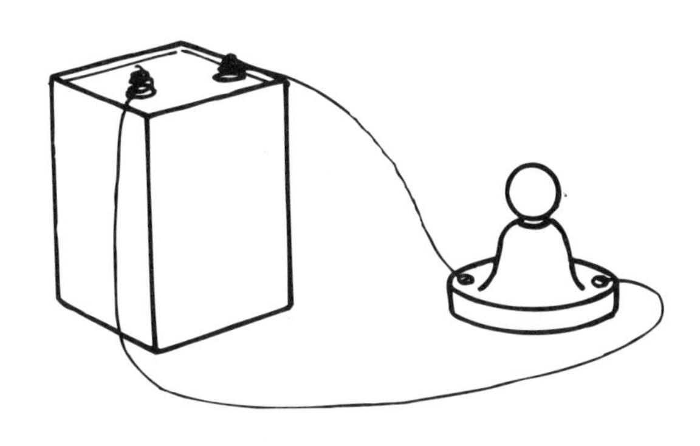

What happens?

____
____
____
____
____
____

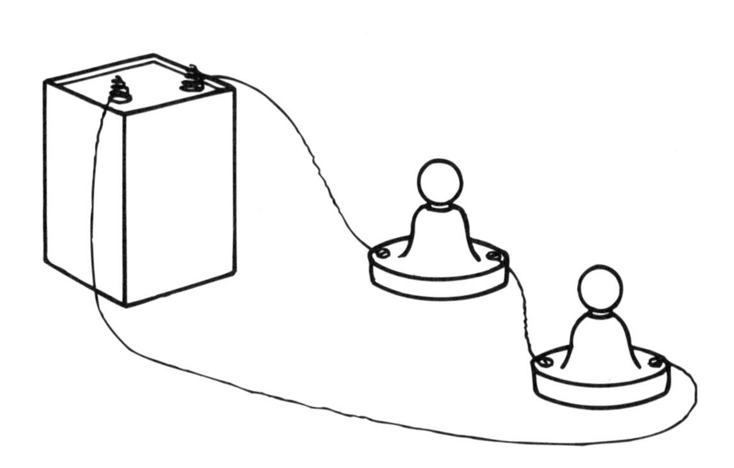

What happens?

____
____
____
____
____
____

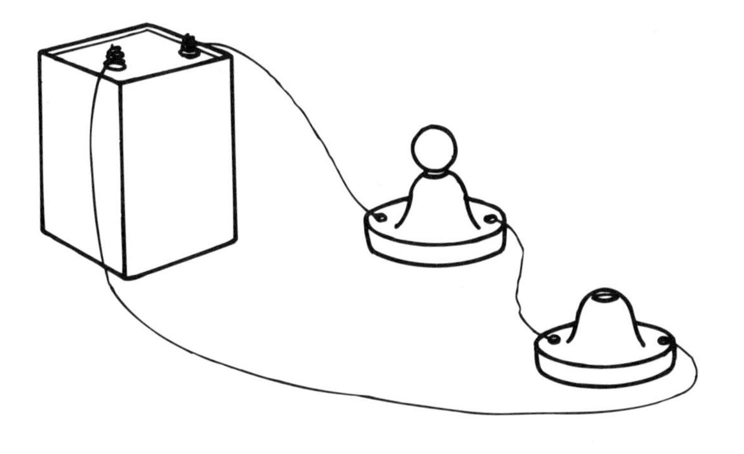

What happens?

____
____
____
____
____
____

Experiment 32 Parallel Circuit

What happens when a circuit runs in loops through more than one light bulb?
Electricity in circuits can produce light.

What You Need:

- Batteries, wires, bulbs, and sockets as in "Series Circuit," on page 70
- One more 1-foot length of wire, ends stripped, for each team
- Activity sheet on page 73 for each student

What to Do

1. Have each team build a circuit that includes both bulbs as in the sketch on the activity sheet. *When both battery terminals are hooked up, what happens?*
2. Leaving all the wires connected as in step #1, unscrew one bulb. *What happens?*

Let's Talk About It

In a *parallel* circuit like this, current runs to two or more appliances (bulbs) in separate loops. Removing one bulb does not interrupt the circuit, so current continues to flow to the other bulb.

Teacher Tips:

Standard diagrams of series and parallel circuits look like this:

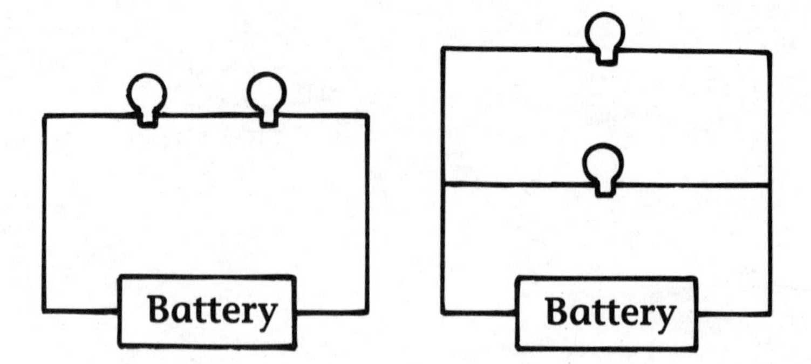

0-7424-2749-8 *Hands-On Physics Experiments*

Name ______________________________

Date ________________

Parallel Circuit

Use two different colored lines to show the current to the two bulbs.

What happens?

____

____

____

____

____

____

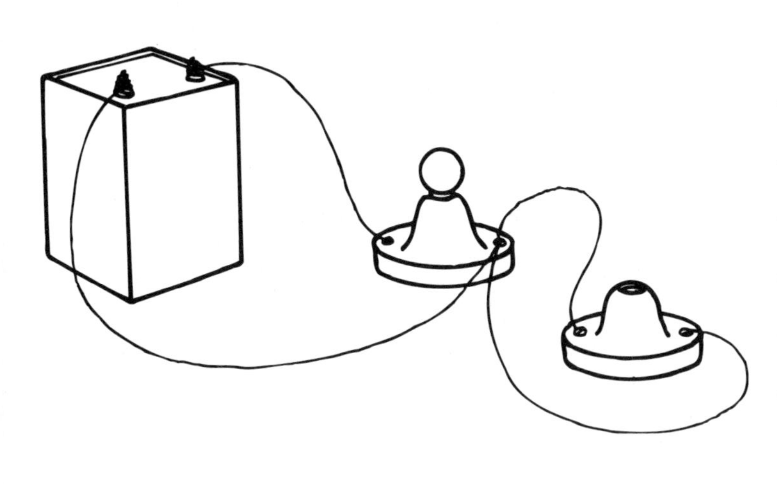

What happens?

____

____

____

____

____

____

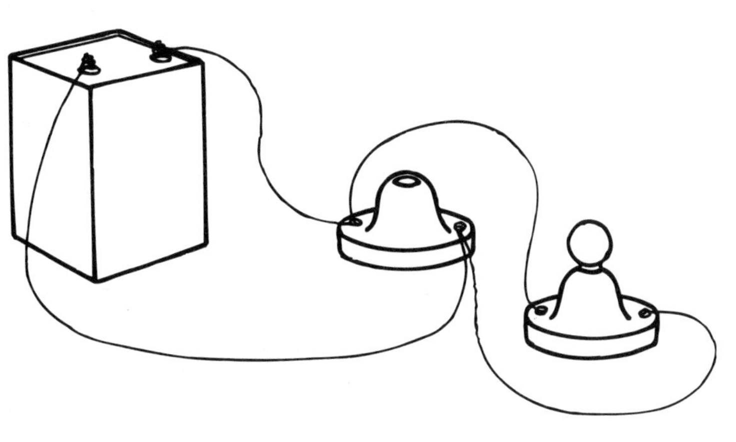

What happens?

____

____

____

____

____

____

0-7424-2749-8 *Hands-On Physics Experiments*

Experiment 33 Making a Magnet

Students use electricity to make a magnet.
Electricity in circuits can produce magnetic effects.

What You Need:

- Six-volt battery for each team
- Copper wire without insulation, about 2 feet for each team
- Iron or steel nail, 2–3 inches long, for each team
- Steel paper clips, one for each team
- Magnet
- Activity sheet on page 75 for each student

What to Do:

1. Have students verify that both the nail and the paper clip will be attracted to a magnet, but that neither one is itself a magnet.
2. Students wrap the copper wire around the nail many times. Keep the wire curling in the same direction, and do not let it overlap itself. It can touch the nail but does not have to. Leave several inches at each end of the wire free.
3. Connect one end of the wire to the battery's positive terminal and the other end to the negative terminal. *Will the nail attract the paper clip now?* Disconnect one end of the wire. *What happens?*

Let's Talk About It:

Electrical current (the flow of electrons) in the coiled wire aligns electrons in the nail, making the nail act like a magnet. When the current stops, the nail's electrons lose their alignment. This is how many large industrial magnets work. They are *electromagnets* that act as magnets only while the current is on.

Name ______________________________

Date ________________

Making a Magnet

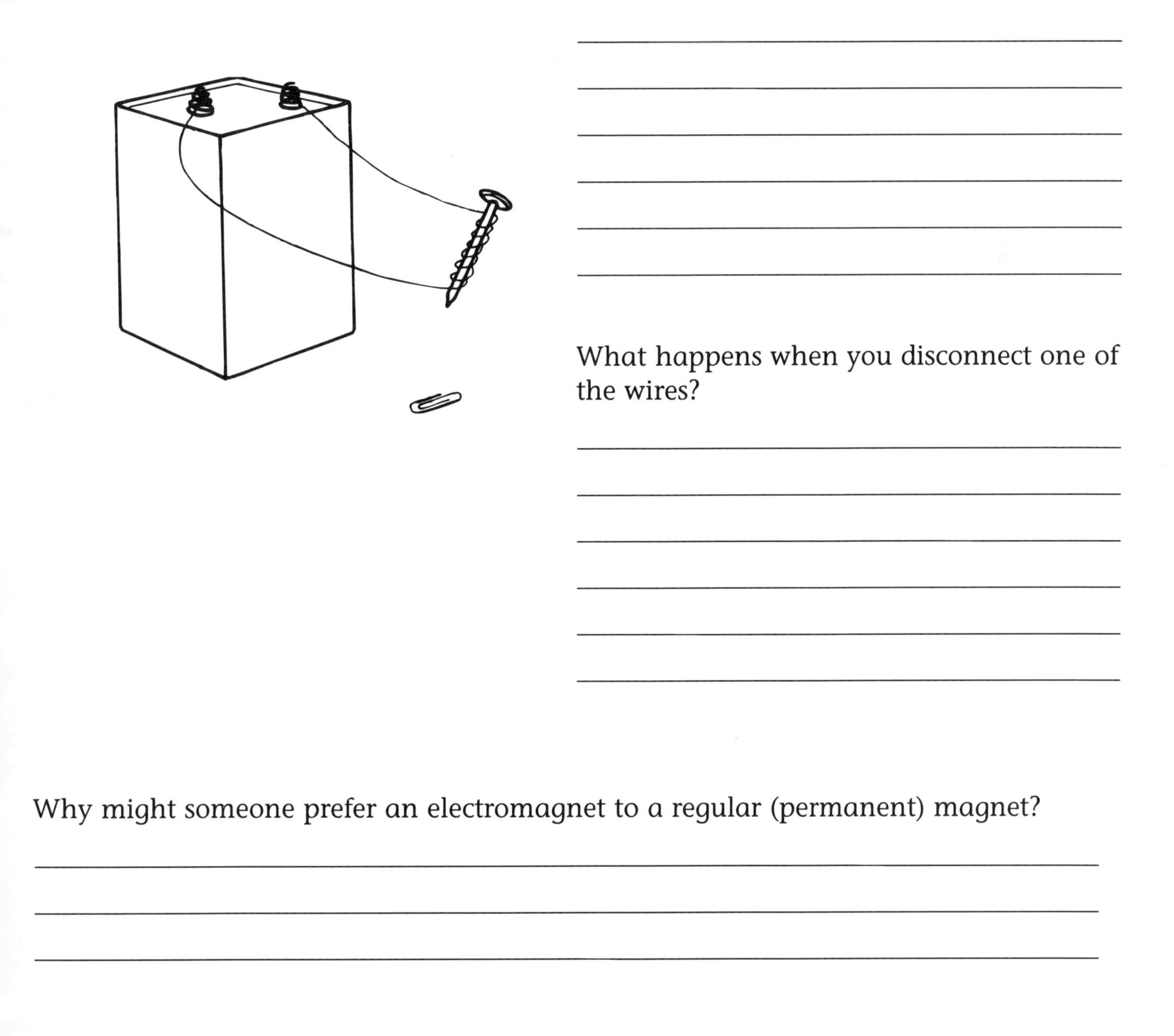

What happens when both ends of the wire are connected to the battery?

____

____

____

____

____

____

What happens when you disconnect one of the wires?

____

____

____

____

____

____

Why might someone prefer an electromagnet to a regular (permanent) magnet?

____

____

____

0-7424-2749-8 *Hands-On Physics Experiments*

Experiment 34 Pulling Through

Do magnets act through solid objects?
Magnetic fields penetrate many materials.

What You Need:

- Bar magnets, one for each team
- Steel paper clips, one for each team
- Assortment of common objects and materials such as paper, cloth, plastic, glass, pottery, and wood
- Rulers, one for each team
- Activity sheet on page 77 for each student

What to Do:

1. Have students find out how far from the paper clip their magnet can be and still attract the paper clip. Call this the magnet's range.
2. Ask: *Do you think the magnet will attract the paper clip if you put something between them?* Have each team choose a material or object to test. Ask: *Will your magnet hold a paper clip through a thin layer of your test material?*
3. If their magnet works through a thin layer of test material, ask: *Will your magnet still work if you use a thicker layer of test material? At what thickness will the magnet no longer hold the paper clip? Is that the same as the range you found in step #1?*

Let's Talk About It:

Magnetic fields work through many materials and objects, even those that do not respond to magnets.

Name ______________________________

Date ________________

What material or object did you test?

____

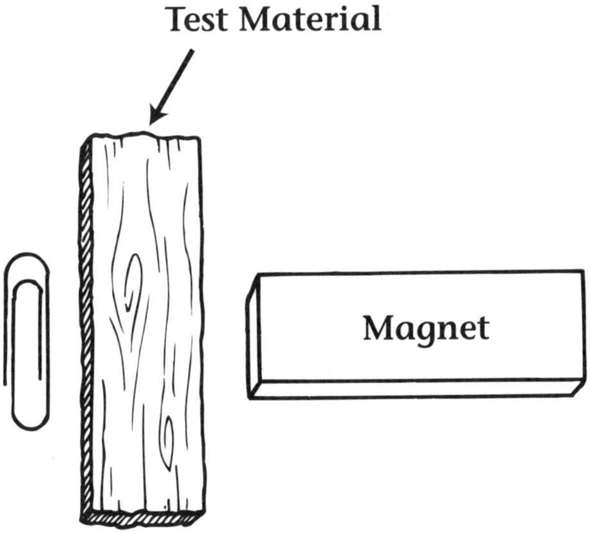

Test Material

Magnet

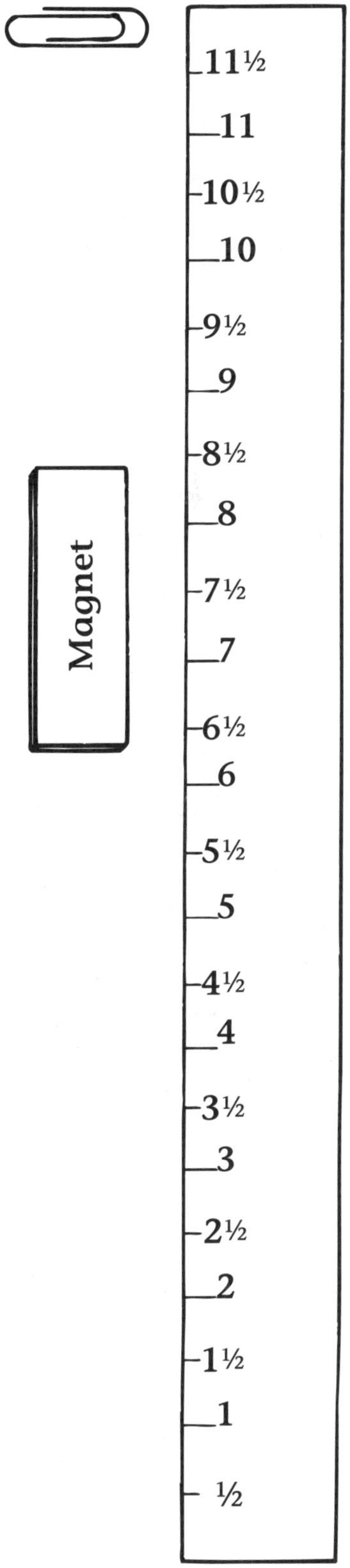

Magnet

11½
11
10½
10
9½
9
8½
8
7½
7
6½
6
5½
5
4½
4
3½
3
2½
2
1½
1
½

Did the magnet work through the test material? ____________

If yes, how thick a layer of test material could the magnet penetrate? ________________

Share your results with the other teams.

Did the test materials affect the magnet's ability to hold a paper clip? ________________

Did all the test materials have the same effect? ______________________________

0-7424-2749-8 *Hands-On Physics Experiments*

Experiment 35 Making a Compass

Students make their own magnetic compass.
Magnets align themselves with Earth's magnetic field.

What You Need:

- Bar magnets, one for each team
- Needles, one for each team
- Thread, a piece 6–8 inches long for each team
- Wooden pencils, one for each team
- Books
- Compass
- Scissors, nails, or other objects attracted by magnets
- Activity sheet on page 79 for each student

What to Do:

1. Ask your students where north is. *What reference points do they use?* Find north with a compass. Label the room's north wall with a red *N*.
2. Have each team magnetize its needle by stroking it 40 times in one direction with one end of their bar magnet. Tie a thread around the middle of the needle, and hang the needle from a pencil. Set the pencil on two stacks of books so the needle hangs down between the stacks. *When the needle stops moving, which way does it point?*
3. Have the students bring a metal object such as a pair of scissors toward their needle. *What happens?*

Let's Talk About It:

Any magnet free to move (as when suspended on a string or floating in a fluid) will orient along the earth's magnetic north-south axis.

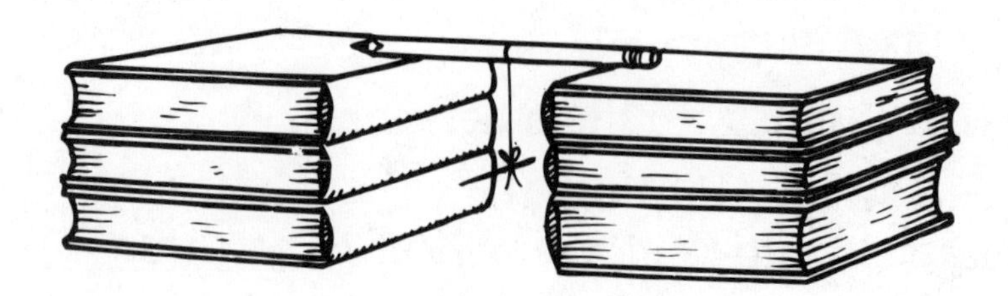

Name ______________________________

Date ________________

Making a Compass

Draw your classroom. Show where north is, according to your teacher's compass. Then show where your needle compass pointed.

Did all the needle compasses in the class point the same direction? ____________

What happened when you moved something metal close to your needle compass?

____

If you were preparing a backpack for a camping trip, would you pack your pocket knife next to your compass? Why or why not?

____

____

____

Further Inquiry Into Light, Heat, Electricity, and Magnetism

Discuss where we use lenses in everyday life. Examples include eyeglasses, binoculars, cameras, flashlights, and automobile headlights. *How do these lenses affect light and the images we see?*

If you can, take a close look at a real rainbow. Record the colors you see, and the order in which they appear. Students interested in learning more about light and color can find out more about wavelengths, prisms, and how the human eye perceives color.

Your students might want to repeat *Light and Heat* (page 62) with *in-between* colors. *How can we use color choices to help us stay cool in hot weather or warm in cold weather?* Discuss this with regard to clothing, automobiles, and houses.

Ask your students why we blow on our hands to keep them warm and blow on soup or hot cocoa to cool it down. *Can they design an experiment to find out if this action actually does what we want it to?*

In *Making a Magnet* (page 74), your students found that their electromagnet lost its magnetic ability as soon as the current was turned off. In *Making a Compass* (page 78), how long did their needle stay magnetized? *Is it possible to destroy a magnet's attracting ability by shaking it? Freezing it? Heating it?*

A compass points toward magnetic north, which is not the same as geographic north. Find Earth's magnetic and geographic poles on a globe. Explain that the geographic poles are along Earth's axis of rotation, while the magnetic poles indicate how Earth's magnetic field is oriented. Interested students can research how that affects the accuracy of compasses.

Metals that respond to magnets can pull a compass needle off line. How do the compasses on metal ships and planes stay accurate? Magnets also affect electronic devices. Try running a magnet over an expendable video. Does it destroy the images?